Genetic Mosaics and Other Essays

The John M. Prather Lectures, 1965

Genetic Mosaics
and Other Essays

by Curt Stern

Harvard University Press

Cambridge, Massachusetts | 1968

Preface

The Prather Lectures for the year 1965 had the purpose of celebrating the centenary of Mendel's presentation of his Experiments on Plant Hybrids before the Brno Natural History Society. The series of three lectures from which this volume is derived dealt with a variety of topics of "classical" and developmental genetics. They were followed by another series of three Prather Lectures given by Dr. Robert S. Edgar on the recent achievements of molecular and microbial genetics.

Whatever unity these lectures might have derives less from their content—although they all deal with genetics—than from the fact that they are the products of a single person. He who is fortunate enough to be a professor at an academic institution should be a scholar who can place the present in its historical perspective, a teacher who can draw together the many threads spun by a host of individual researchers, and an investigator who himself contributes some new facts and ideas. It is most exceptional to find these various tasks equally well fulfilled by the same person, and the present instance is no such exception. Nevertheless, an attempt has been made to respond to the different demands outlined above. The first lecture in this series, "Mendel and Human Genetics," places the history of human genetics in the perspective of Mendel's influence. The second lecture, "Genetic Mosaics in Animals and Man," provided me with an opportunity to look at unique, curious creatures, to consider them from modern viewpoints, and to describe the surprising recent discoveries of genetic mosaicism as a normal property of organisms. This lecture has been much enlarged in its printed version, but even in this form it is no more than a selection from an abundance of material. The last lecture, "Developmental Genetics of Pattern," reviews work that has occupied my associates, students, and myself for a great part of the last fifteen years.

A short "sermon" entitled "Thoughts on Research" was read at the conclusion of the three Prather Lectures. It was written at the suggestion of my wife Evelyn, in order to convey to the hearers some personal opinions and feelings. These few pages are reprinted as the last part of the book.

The welcome accorded to us at Harvard University created an atmosphere of warmth and empathy between the visitors and the audiences of colleagues, both students and faculty. It was a wonderful experience. I wish to express my gratitude to the Committee for the Prather Lectures, particularly its Chairman, Professor Paul C. Mangelsdorf, for the great opportunity which their invitation opened for me.

Curt Stern

University of California
Berkeley
February, 1967

Acknowledgments

I have to thank various authors and publishers for the use of published or original illustrations. Individual acknowledgments are made in the body of the book. The first lecture, "Mendel and Human Genetics," reprinted here in a slightly changed version, first appeared in the *Proceedings of the American Philosophical Society* (1965), and the last section, "Thoughts on Research," has been reprinted from *Science* (1965). Much of the work discussed in the third lecture of the book has been supported by National Science Foundation Grants No. 18057 and GB 2532.

For excellent collaboration in the preparation of this book I am grateful to Eva R. Sherwood.

Contents

FIGURES

Genetic Mosaics and Other Essays

Mendel and Human Genetics

Mendel's clue has shewn the way into a realm of nature which
for surprising novelty and adventure is hardly to be excelled.
It is no hyperbolical figure that I use when I speak of Mendelian
discovery leading us into a new world, the very existence of
which was unsuspected before.   —William Bateson

Mendel left no records concerning studies in human genetics.
His biographer Iltis, however, states that Mendel had shown
persistently great interest in hereditary phenomena in man and
in anthropological and medical problems. He studied the
records of old Brünn families for aspects of inheritance and,
among his own kindred, made careful observations of peculiarities in hair form, hair color, and body size in successive
generations. Regularly he took measures of body dimensions
of his growing nephews and he often attended autopsies in a
hospital. Had Mendel's time not gradually been completely
absorbed by his duties as abbot of his monastery, it might be
surmised that he would have reported on his observations on
man, expanded them, and, with the insight gained from his
experiments on peas, been able to clarify problems of human
inheritance in such a way as to be counted a direct founder
of human genetics. But it hardly changes the situation that
Mendel did not make a specific contribution to that field. His
basic discoveries in plants could be applied to man without
requiring new insights.

Human genetics had a slow growth. This has been ascribed
to the difficulties which man with his long life span, his small
families, and the absence of scientifically controlled matings

1

offers to genetic analysis. It has also been ascribed to the dampening effect on bona fide research which resulted from class and race prejudice within the eugenics movement. Undoubtedly both of these factors had some influence, but perhaps most important was the fact that for decades the best minds working in genetics were interested in the general phenomena of inheritance, not in their expression in specific species. "If you want to study the genetics of rabbits," it was said informally some forty years ago, "study rabbits. If you want to study genetics, study Drosophila!" For a still earlier period replace the word Drosophila by pea (Pisum), fowl, sweet pea (Lathyrus), or gypsy moth (Lymantria), and by bread mold (Neurospora), colon bacterium (Escherichia), or bacteriophage for a subsequent one, and it becomes understandable why human genetics remained peripheral to the center of genetic advance.

Beyond such influences on the development of a field, its history may depend on extrinsic phenomena, such as the state which related areas have attained. Mendel's paper can serve as an illustration. The reason that it remained without influence for thirty-five years and then, on its "rediscovery," was immediately recognized in its importance, depended primarily on what had happened during the intervening period. This period included the discovery of chromosomes and their behavior in cell division and gametogenesis, an intensive study of biological variation, and the formulation, by August Weismann, of a conceptual framework for a theory of heredity, development, and evolution. The time was ripe for Mendelism.

One additional element should be mentioned which influences the growth of knowledge. This is the limitation of individuals and communication between individuals which leads them to recognize only slowly the significance of new findings, whether made by others or by themselves. The development of human genetics furnishes examples of such delays, as will be shown in the following pages.

There are many interweaving threads in human genetics. A few main strands will be followed here in artificial separation from one another. The discussion will halt at the time when human genetics became of age, at the end of the period between the two World Wars.

Pedigree Analysis. Careful descriptions of pedigrees and the establishment of empirical rules of inheritance of specific traits preceded the recognition of Mendel's work (Fig. 1, center). Long ago the Talmud gave evidence of knowledge concerning the mode of transmission of hemophilia (M. Fischer 1932), and Nasse, in 1820, gave a specific formulation of it. Maupertuis and Réaumur, in the middle of the eighteenth century, each described in admirable detail a pedigree of polydactyly (Glass 1959). The discovery of red-green color blindness in an English boy (Whisson 1779) was accompanied by the realization of its occurrence in other members of his family and of its hereditary

FIG. 1. Trends in the history of human genetics: pedigree analysis, consanguinity, and mutation. In this figure as well as in the similar Figs. 3 and 4, the selection of names of investigators does not lack arbitrariness. While the inclusion of persons can probably be defended in most cases, absence of names can often be excused only by the attempt not to overload these schematic presentations. Names in parentheses refer to individuals not directly concerned with human studies. The word *Treasury* refers to a specific serial publication (see Pearson 1912–).

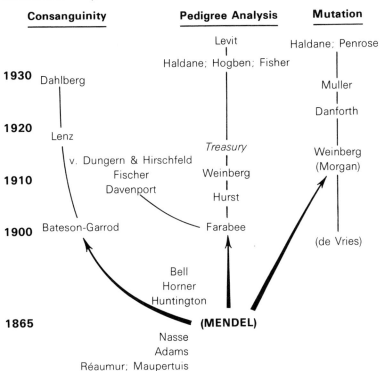

nature. This was followed by other studies of "Daltonian" color blindness and, one hundred years later, by Horner's codification of the empirical facts of transmission. In 1814 the physician Joseph Adams wrote a penetrating *Treatise on the Supposed Hereditary Properties of Diseases,* in which the modern reader can discern many of the features of both general and specifically human genetics now familiar to us. Sedgwick, in 1861, assembled extensive data "On Sexual Limitation in Hereditary Disease," and Huntington, in 1872, accurately described the mode of transmission of the hereditary chorea which now carries his name. In an impressive memoir, "Upon the Formation of a Deaf Variety of the Human Race," Alexander Graham Bell, in 1883, showed the hereditary basis of many instances of deafness and emphasized the fact of preferential marriages between deaf mutes.

All these studies, however, remained separate and mostly without further consequences. Even when the rules of inheritance of a trait were clearly described, no underlying causes were recognized, so that the level of the discussions rose only slightly above that of casuistics. This became different as soon as Mendel's work was rediscovered. When Garrod began to study the biochemistry of alkaptonuria and became aware of its familial occurrence, Bateson suggested that it might well be a recessive Mendelian trait (1902). A year later Farabee (1903a) described a Negro kindred with several occurrences of albinism as evidence for the recessive nature of this trait. During the winter of 1901–1902, at Harvard University, Farabee had attended Castle's lectures on inheritance in which Castle interpreted albinism in mice as being due to a recessive allele. In 1903 (b) Farabee filed his doctoral dissertation, which contains a description of a large Pennsylvania kindred with brachydactyly and demonstrates convincingly that the trait followed the transmission of a dominant gene (or "character" as it still was called).

Ever since, in a steady stream, innumerable pedigrees have been published, largely concerned with abnormal and medically significant traits. Variations within the normal range were likewise interpreted in Mendelian manner, eye and hair colors being the first two examples (Davenport and Davenport 1907, 1910; Hurst 1908). A third example concerns the human A-B-O blood groups which were discovered by Landsteiner (1900, 1901). Their inherited nature was recognized by von Dungern and Hirschfeld, and a two gene-pair hypothesis was proposed. We shall return to this topic. Mendelian segregation also became apparent in racial crosses, as for various anthropological traits in the Rehoboth hybrids between Caucasians and Hottentots in South Africa (E. Fischer 1913), and, for skin color, in the Negro-Caucasian hybrids of Jamaica (Davenport 1913). This was of general significance since the belief in a non-Mendelian "blending" inheritance in racial mixtures had been particularly strong.

The recording of a single pedigree or a few pedigrees for a given trait was expanded into monographic treatments of all its known incidences. Pearson and others' monumental *Monograph on Albinism* (1911–1913) and the series of many volumes entitled *The Treasury of Human Inheritance*, which was founded by him in 1912 and still continues publication, as well as Cockayne's *Inherited Abnormalities of the Skin and its Appendages*, furnished abundant data not only for the conclusions of their authors but also for independent investigators in need of facts.

The enthusiasm of many workers was both a boon and a danger in the early days of genetics. Hurst, for instance, the student of eye-color inheritance, was "a tireless worker and full of ideas, but over-apt to find the 3:1 ratio in everything he touched" (Punnett 1950). Only gradually did a truly critical attitude toward pedigree studies evolve. The first one who saw the methodological problems posed by analyses of pedigrees was the physician Wilhelm Weinberg. Beginning in 1908 with a paper on the demonstration of inheritance in man, and particularly in a publication of 1912 on methods and sources of error in studies directed toward Mendelian proportions in man (1912a), he devised means of correcting for various types of ascertainment. The fact which had concerned Bateson (1909) and for which Garrod showed an intuitive though not explicit understanding, namely, that the fraction of albinos in segregating sibships from normal parents is higher than the fraction ¼ expected from $Aa \times Aa$ crosses, became comprehensible as the necessary deviation from Mendelian expectation when it remained uncorrected for biases introduced by family selection. Little further progress in this area was made for twenty years; then Hogben (1931), Haldane (1932a), Fisher (1934), and others began to apply their rigorous minds to the problems of pedigree analysis. During this period also it was recognized generally that the manifestation of many genotypes varies from person to person. The bearing of incomplete penetrance on the problem of dominance in man underwent an important analysis by Levit (1936).

The knowledge of Mendelism early made possible a new interpretation of the effects of consanguinity (Fig. 1, left). This age-long problem had been actively followed in the nineteenth century, with the pioneering inquiry by Bemiss (1858) deserving specific mention. Again, the facts gained remained without a rational foundation. When, however, Garrod (1901) noticed that in three out of four sibships containing alkaptonuric individuals the parents were first cousins or otherwise closely related, Bateson (1902) furnished the explanation. He reasoned that a person who carries a rare recessive gene would have a low chance to marry an unrelated person who also carried this rare gene, but a high chance, in case of consanguinity, to have his spouse be a carrier who had inherited the gene from a common ancestor. It seems to have occurred to no one to provide a quantitative expression for this relationship until Lenz's article in 1919. Then, beginning in 1927, Dahlberg refined the mathematical treatment of consanguinity which in recent years has become a central issue in the study of the genetic load of populations.

The problems of this load will not be a topic of this account, but a brief reference is indicated to the mutations which compose a highly important part of it (Fig. 1, right). Mendel did not make a direct contribution to the problem of the origin of different varieties of a gene, of its alleles. On the other hand, Hugo de Vries was led to his rediscovery of Mendel's paper by his earlier studies of genetic changes in the evening primrose. T. H. Morgan, of course, who first discovered as well as analyzed mutations in Drosophila, came to genetics via Mendel and de Vries. The epochal success of his former student, H. J. Muller, in "Artificial Transmutation of the Gene" (1927) by means of X-rays, was used by him immediately to call attention to the possibilities of radiation-induced damage to human genes. It was not the first time that mutations in human genes had been considered by modern investigators. Weinberg, in 1912 (*b*), noted the tendency of last-born children to be more frequently affected by dwarfism than would be expected by chance. He suggested that, if more exact analysis should indeed show this to be the case, this would speak for mutation from normal to dwarfness increasing in frequency with age of the parents. Danforth, in 1923, actually estimated human mutation rates, assuming an equilibrium between mutational input and selective outgo of unfavorable genes. His pioneering paper remained without consequences and the same method had to be re-invented, in 1935, by Haldane and by Penrose (see Gunther and Penrose).

Population Genetics. Mendel himself was the first who, on the basis of his particulate theory of inheritance, attacked a problem not only of the genetics of individuals and their progeny, but also of a whole population. After having established the 1:2:1 ratio for homozygotes of one kind: heterozygotes: homozygotes of the alternative kind in the F_2 generation of his crosses, he asked what ratios would be found in the further generations of the F_2 population. Peas are self-fertilizing plants, and Mendel found that in this instance the proportions of the three genotypes change from generation to generation. He obtained an algebraic expression for this change, yielding the proportions $2^n-1:2:2^n-1$ where n is the number of generations beginning with the F_2 (Fig. 2).

FIG. 2. Mendel's analysis of a problem in population genetics. From pages 37–38 of the facsimile of Mendel's manuscript of his "Versuche über Pflanzen-Hybriden," reprinted in L. Gedda, *Novant'anni delle leggi Mendeliane* (Rome: Instituto "Gregorio Mendel," 1956).

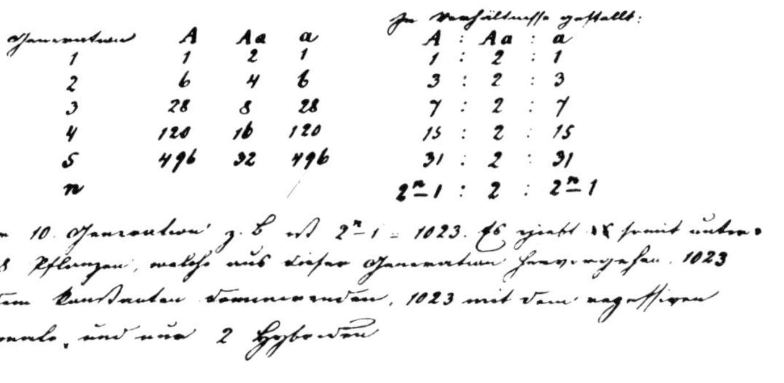

	A	Aa	a	A	:	Aa	:	a
1	1	2	1	1	:	2	:	1
2	6	4	6	3	:	2	:	3
3	28	8	28	7	:	2	:	7
4	120	16	120	15	:	2	:	15
5	496	32	496	31	:	2	:	31
n				2^n-1	:	2	:	2^n-1

Most organisms do not reproduce by selfing, a fact which is obvious for species with separate sexes. It was natural, therefore, that the question arose concerning the proportions of genotypes in various generations of crossbreeding forms, but the way in which this came about and the slow steps by which it was answered make a fascinating chapter of genetics (Fig. 3, left).

Before 1900, begun by Galton and deepened by Karl Pearson, a biometric school had developed in England which formulated laws of inheritance of a statistical nature. Basing their work primarily on measurements of human stature, the biometricians determined correlation coefficients between groups of parents and children, and between other groups of related individuals.

FIG. 3. Trends in the history of human genetics: population genetics and nature-nurture.

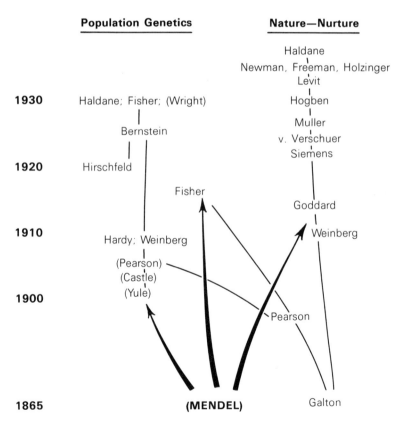

Galton (1897) derived from his data a "Law of Ancestral Heredity": "The two parents contribute between them, on the average one-half . . . of the total heritage of the offspring; the four grandparents, one-quarter . . .; the eight great-grandparents, one-eighth . . ., and so on." Pearson (1904) modified Galton's specific conclusions but upheld its basic tenets. It was the assumption of genetic contributions of the pre-parental ancestors which disturbed the Mendelians. Mendel's law of segregation centered on the purity of gametes. A homozygous recessive parent, even if derived from two heterozygotes, would transmit only his own recessive genes, uncontaminated with dominant ancestral influences. Likewise, a homozygous dominant parent with earlier recessives-bearing parentage would transmit only his own dominant genes, free from recessive ancestral contamination. Bateson insisted that Mendel's law of segregation and the law of "ancestral heredity" could not both be applicable to the same class of cases. The biometricians, on the other side, either felt that there was no inconsistency between the two, or that Mendelian genetics was not applicable generally. The basic conflict between Galton's and Mendel's laws led to a violent controversy in which the standard bearers were W. F. R. Weldon for the biometricians and Bateson for the Mendelians.

In 1902 the statistician Yule published a discussion on Mendel's laws in which he took Weldon's side, though he was "inclined to agree with Mr. Bateson as to the *possibly* very high importance . . . of Mendelian phenomena." In his paper he correctly derived the fact that an F_2 generation would preserve its 1:2:1 ratio in all later generations, provided that random mating between all individuals took place. This was an important insight, but it was lost in what followed in the same and subsequent paragraphs. Yule was interested in the frequency of dominants in successive generations among the offspring of the dominants mated at random to other members of their generation. He found that there would be an increase of the dominants from 75 percent in F_2 to 83.3 percent in F_3, to 85.0 percent in F_4, approaching a limiting value of 85.355339 in a few further generations. Yule regarded this increase in the frequency of dominants with increasing numbers of generations of known dominant ancestors as demonstrating "as nicely as could be desired" one of the chief properties of the law of ancestral heredity, namely the influence of pre-parental generations on the offspring. He had to admit, however, that the law does not hold for recessive attributes, since recessive homozygotes, regardless of their ancestry, can only produce recessive germ cells.

Yule's treatment of the dominants was based on a limited number of all possible crosses within the population at large. He did not state very clearly what he was doing and his wording easily produced the impression that he was trying to derive the proportion of dominants in successive generations after selection against the homozygous recessives. Castle, the American pioneer in Mendelism, misinterpreted Yule in this way, and was stimulated to a treatment of his own (1903). He established once more the constancy of the 1:2:1 ratio in any panmictic generation following the F_2 and then proceeded to demonstrate that selection against the recessives would not lead to a limiting value of dominants but to a continuous increase from 75 to 88.8 for the first generation following selection of the dominants, to 93.7 for the second generation, to 96 percent for the third, gradually approaching 100 percent with continuous selection in each generation. Castle went beyond this. He found that whatever ratio of the genotypes had been reached during the generations of selection would remain constant in future generations once selection had been discontinued. "In general, as soon as selection is arrested the race remains stable at the degree of purity then retained . . ." Thus, he had discovered an equilibrium law not only for the 1:2:1 ratio but for the infinitely large class of all ratios resulting from selection against recessives. With this successful defense of Mendelism against Yule's presumed position, Castle returned to his breeding experiments, little aware of the gold which he had found in the still undefined area of population genetics. His paper remained largely unnoticed and even Yule seems not to have responded to it in printed form. When, twelve years later, M. T. J. Norton provided Punnett with a table on the effect of selection against Mendelian genes (Punnett 1915) and, further two years later, Punnett, with help from the mathematician Hardy, wrote a note (1917) in which he showed the slow progress to be expected from the elimination from reproduction of supposedly simple homozygous recessive feebleminded individuals, they essentially followed Castle's steps (1903) but did not know that they had a predecessor.

Within a year after Castle, and obviously without knowing of his findings, Pearson himself proved the stability of the 1:2:1 ratio in a panmictic population. He also derived the results of panmixis when more than one pair of alleles were involved but always still based on an initial F_2 generation.

In modern terms, Pearson considered only the case of equal frequency of the two contrasting alleles A_1 and A_2 (or A and a), while Yule and Castle included a series of selected frequencies. It is strange indeed that the outstanding biometrician, Pearson, did not see the need for expanding the study to all frequencies.

The matter rested until 1908. On February 28 of that year Punnett, one of the closest associates of Bateson, gave a lecture before the Royal Society of Medicine in London entitled, "Mendelism in Relation to Disease" (Punnett 1908). In the discussion which followed the lecture, and which is reported in the printed *Proceedings,* Yule referred to some of the figures illustrating the Mendelian cases, which puzzled him very much.

Assuming that brown . . . eye-colour was dominant over blue, if matings of persons of different eye-colours were random . . . it was to be expected that in the population there would be three persons with brown eyes to one with blue; but that was not so . . . The same applied to the examples of brachydactyly. The author said brachydactyly was dominant. In the course of time one would then expect . . . to get three brachydactylous persons to one normal, but that was not so.

Now it was occasion for Punnett to be puzzled. He reworded, somewhat inaccurately, Yule's comments as "Mr. Yule wondered why the nation was not slowly becoming brown-eyed and brachydactylous . . ." and replied, "So it might be for all he knew, but this made no difference to the mode of transmission of eye-colour or brachydactyly." Punnett, however, did not feel content with his own comment. On his return to Cambridge he at once sought out G. H. Hardy, whom he knew well, for they acted as joint secretaries to the Committee for the Retention of Greek in the Previous Examination and also used to play cricket together (Punnett 1950).

Knowing that Hardy had not the slightest interest in genetics I put my problem to him as a mathematical one. He replied that it was quite simple and soon handed me the now well-known formula $pr = q^2$.[1] Naturally pleased at getting so neat and prompt an answer I promised him that it should be known as "Hardy's Law"—a promise fulfilled in the next edition of my *Mendelism*.

The essence of Hardy's finding is the constancy of the distribution of the three genotypes after the second generation whatever the values of p, q, and r may be, that is, whatever the frequencies of the two alleles may be. Specifically, Hardy showed through two examples that the proportion of brachydactylous persons, if the trait is dominant, will have no tendency whatever to increase, and if it were recessive, would have no tendency to decrease. It may be added that, while Hardy pointed out that he had considered only the very simplest hypothesis possible, he had actually treated, for the first time, the problem of genetic drift which was to become so important in Sewall Wright's later work. There is a postscript to Hardy's one-page note according to which Yule would accept its substance "as a satisfactory answer to the difficulty that he raised." It has been said that Nature yields answers only to correctly formulated questions. Hardy's solution to Yule's difficulty shows that wrong questions may sometimes be fruitful also.

I have wondered occasionally whether the statistician Yule's original question was asked seriously or rather with tongue-in-cheek in order to embarrass the Mendelian lecturer. Apparently this suspicion is quite unjustified. The question rather shows how a distinguished statistician could miss the general concept of allele frequencies which appeared so obvious to Hardy who could find nothing more subtle to apply to it than "a little mathematics of the multiplication-table type." Hardy's note was published in *Science*. "The reason why it appeared . . . [there] is that *Nature* at that time was extremely hostile and refused to publish anything tainted with Mendelism."[2]

Hardy's Law remained known under this designation until 1943 when it was realized that, independent of Hardy and

---

1. "Where p, $2q$ and r are the proportions of AA, Aa, and aa individuals in the population varying for the A-a difference." See footnote p. 9 in Punnett 1950.
2. R. C. Punnett, Feb. 5, 1950, in letter to the author.

indeed at least six weeks prior to Hardy's involvement in genetics, Weinberg had presented the equivalent formula before the Society for Natural History in Stuttgart (Stern 1943). The publication of his paper (Weinberg 1908) also preceded that of Hardy's. Weinberg came to it as a biologist and physician who had embarked on a wide-ranging mathematical treatment of problems of human genetics. His approach was less abstract than that of the mathematician Hardy. The idea of allele frequencies is not explicitly expressed by either Weinberg or Hardy, since both start with frequencies of genotypes. Instead of Hardy's three interrelated parameters p, q, and r, Weinberg uses only two, m and n, which represent the frequencies of the two initial homozygous populations AA and aa thus actually allele frequencies. He expresses the result of panmixis for the first time in the now familiar form:

$$m^2AA + 2mnAB + n^2BB \quad \text{(where } A \text{ and } B \text{ are alleles).}$$

The names of both the discoverers are now attached to the population formula: The Hardy-Weinberg Law.

In 1909 Weinberg generalized the theorem in terms valid for multiple alleles, and investigated polyhybrid populations in which he recognized their essentially different method of attaining equilibrium. Since that time the concept of allele frequencies and the formula for equilibrium in case of panmixis in Mendelian populations have been the foundation for population genetics in general.

A very impressive application of an expanded Hardy-Weinberg formula was made by Bernstein (1924, 1925). This mathematician had earlier become interested in human genetics and had interpreted population data on variations in singing voice and direction of hair whorl as found in different populations in terms of allelic differences of single pairs of genes. His evidence consisted in a fit of the proportions of phenotypes to the $p^2:2pq:q^2$ expectation (where p and q correspond to Weinberg's m and n, and not to Hardy's terms). Bernstein then turned to a population genetic analysis of the frequencies of the four blood-group types O, A, B, and AB. Numerous records of racially variant blood-group frequencies were available, beginning with the discovery of this phenomenon by L. and H. Hirschfeld (1919).

As noted earlier in this essay, a genetic interpretation of the blood-group variations had been given. It assumed the existence of two pairs of alleles, A-a and B-b. When Bernstein compared the expectations for the blood-group frequencies according to the dihybrid Hardy-Weinberg formula with the observed proportions, he found significant and consistent differences. He concluded that the blood groups were not inherited in the hitherto accepted fashion and searched for a different interpretation. He thought of a triple-allelic system of a single locus, applied the appropriate equilibrium formula, and found excellent agreement between expectation and observation in frequencies in diverse populations. This Bernstein regarded as proof of his multiple-allele theory, notwithstanding the fact that the limited amount of published pedigree data contained cases not in conformity with expectation from the theory. Later the apparent exceptions could be ascribed to various sources of error, and Bernstein's interpretation has long been fully established.

I might record here an incident which illuminates the newness and the power of the population genetics approach in the analysis of modes of inheritance. In the spring of 1933, Bernstein gave a seminar at the California Institute of Technology in which he reported on his mathematical approaches to human genetics with emphasis on blood groups. In the discussion, T. H. Morgan commented in his quizzical manner that Bernstein's approach was interesting, but could the solution not just as well have been obtained from pedigree analysis? Bernstein, as I remember it, replied that it could but that it wasn't!

Weinberg had early shown (1909, 1910) that the biometricians' Law of Ancestral Heredity was fully compatible with a Mendelian interpretation if it included significant contributions to variance of nongenetic type as well as considerations of polygenic inheritance. The latter had become famous since Nilsson-Ehle's analysis of continuous variation in the color of wheat grains, but this concept of multifactorial inheritance had been fully conceived by Mendel, who had used it to provide a tentative explanation for continuous variation in flower color as observed in the F_2 crosses between two species of beans. In countries other than England the controversy between the biometricians and the Mendelians had never played an important role. Weinberg's demonstration, therefore, made no impression on non-British geneticists and was apparently missed by the British school. Only in 1918 did an analysis by R. A. Fisher lead to a generally accepted Mendelian interpretation of the pre-Mendelian findings of Galton, Pearson, and their school. The subsequent rise of higher population genetics, beginning with the work of Haldane, Wright, and Fisher in the 1920's, cannot be a topic of this survey, even though it has exerted a fundamental influence on human genetics.

Nature and Nurture. Mendel was aware of variability in gene expression due to environmental conditions, specifically as regards the flowering time of his plants. He also noted that some characters do not permit a sharp and certain separation but exhibit continuous variation. For his main studies he deliberately selected such traits as appeared in the plants "clearly and definitely," a virtue in experiments whose purpose was to clarify the basic ways of hereditary transmission. Unfortunately, it led some of his intellectual descendants to reverse the argument. They would classify in alternative ways what was not clearly and definitely distinct, and then account for the alternatives in terms of single gene pairs. Among the many examples of this procedure, one of the best known was Goddard's attempt to explain all feeblemindedness as due to a single homozygous recessive gene (Fig. 3, right). We now know that such types of feeblemindedness do indeed exist, but that they make up a small minority of all afflictions of this type. Of the majority, some are caused by birth injuries, others are due to complex and poorly understood polygenic types of inheritance, still others by perhaps even more complex and poorly understood social influences which help to push an individual below the arbitrary line separating low normality from "feeblemindedness" (or "mental retardation" as contemporary nomenclature prefers to call it). Perhaps most important, there are complex and poorly understood interactions between genetic and nongenetic agents which assign a person to his place in the wide range of mental performance encountered in human populations.

It was the nature-nurture problem which played such an important role in Galton's early creation of human genetics in pre-Mendelian terms. In this pioneering work the influence of the environment on many traits, particularly those involving mental abilities and achievement, was clearly realized in a general way. On the other side, Galton (1908) concluded that "when the nurtures of the persons compared were not exceedingly different," under these circumstances "the evidence was overwhelming that the power of nature was far stronger than that of nurture."[3] On the basis of this conviction, Galton founded the eugenics movement, which attracted many well-meaning but often class-prejudiced adherents. When Weinberg began his population genetics work, he became the first investigator who partitioned the total variance of observed phenotypes into genetic and environmental portions and therewith reconciled the two independent doctrines in genetics originated by Mendel and by Galton (Weinberg 1909, 1910).

Galton made use of the similarities and differences between twins to judge the relative importance of hereditary versus environmental agents in human variation. In the nineteen-twenties research on twins, begun by Siemens and von Verschuer, became widely practiced for the study of normal and particularly of medically significant traits. The interpretation of the findings of the investigators who studied twins depended on the fact that identical twins have identical genotypes, while nonidentical twins are genetically different. Because twins of both kinds are usually raised in the same home, it was often assumed that the environmental influences on pairs of identical twins were of the same degree of similarity as those on pairs of nonidenticals. Additional parameters were desirable such as those provided by cases where twins had been reared apart in different homes, or where children had been adopted away from their biological parents, or where groups of children from different parents had been raised in the relatively uniform environment of an orphanage. Muller (1925) was the first to describe in detail identical twins reared apart, with emphasis on mental attributes; Barbara Burks compared foster-parent–foster-child resemblance in mental scores and achievement with that of true-parent-true-child resemblance; and Evelyn Lawrence studied the children in an orphanage. This type of analy-

3. A penetrating treatment of the nature-nurture problem with special reference to Galton is given in Weinstein 1933.

sis culminated in *Twins: A Study of Heredity and Environment,* a joint work by Newman, the biologist, Freeman, the psychologist, and Holzinger, the statistician (1937).

While such fundamental studies went on, the unscientific literature on class and race genetics, with its ultimately tragic consequences, exerted an inhibitory influence on wider participation in research in human genetics. Some of these interrelations have recently been traced by Dunn and Haller. But while many biologists stayed away from human genetics, other outstanding investigators with deep interest in the social consequences of science entered the field as scientists. They dissected unproven or false opinions, and made contributions of their own which strengthened human genetics as an objective discipline. As eminent examples of influential critics we may cite Hogben with his book on *Nature and Nurture* (1933) and Haldane with *Heredity and Politics* (1938), and both, in their research reports, as creators of new tools and concepts.

Cytogenetics. It was historically impossible for Mendel to have been directly involved in cytogenetic problems. When he wrote his paper in 1865 chromosomes had not yet been discovered. He did refer to the factors which were responsible for the traits he observed as cell elements. He did establish the basic fact—although he communicated it only in two letters to Nägeli (3 July and 27 September, 1870)—that a single pollen grain is sufficient to fertilize an egg cell. And he carried out crosses involving plants with separate sexes—again reported in a letter only (27 September 1870)—whose results suggested to him that perhaps sex is inherited in a way similar to that of other segregating characters.

If Mendel himself had nothing to do with cytogenetics, Mendelism, of course, was one of its two pillars. Very briefly, therefore, the earlier history of human cytogenetics will be sketched here (Fig. 4, left). It was Flemming, the pioneer student of mitosis, who first estimated the number of chromosomes in human tissue cells and believed it to be close to 24. This was corrected by von Winiwarter (1912), who counted 47 chromosomes in male and 48 in female cells. Painter, eleven years later, confirmed von Winiwarter's findings except that he saw 48 chromosomes in both sexes. It is well known now that the true somatic chromosome number in man is 46, a fact established by Tjio and Levan only in 1956, well beyond the period which our survey covers.

The relation of specific chromosomes to sex determination had first been discovered in insects. It was assumed that sex chromosomes also exist in man. The uneven number of chromosomes seen in cells of the human male by von Winiwarter indicated an XX = ♀, XO = ♂ mechanism, but Painter demonstrated clearly that while women have indeed two X-chromosomes, men have one X- and one Y-chromosome. Morgan as well as Wilson recognized in 1911 that the transmission of red-green color blindness could be understood on the basis of color vision genes located in the X-chromosome. In 1922, Castle and Enriques suggested that the Y-chromosome

FIG. 4. Trends in the history of human genetics: cytogenetics and biochemical and molecular genetics.

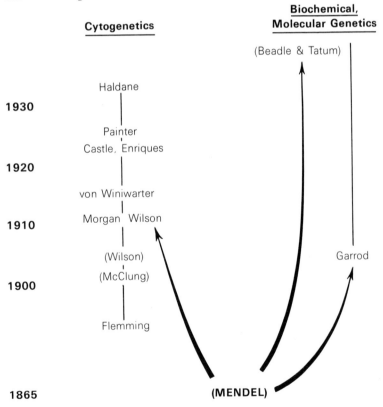

was the bearer of a gene for webbed toes whose transmission in a certain family seemed to follow the male line exclusively. But while the nature of the X-chromosome as carrier of numerous X-linked genes has been firmly established, the existence of Y-linked inheritance apart from determination of male sex is still under discussion.

Chromosomal abnormalities as causes of unusual transmission of human traits were first suspected in a family with a defect of color vision independently by T. H. Morgan and several other geneticists, among them Haldane (1932b), who prophetically advocated in such cases the chromosomal study of leukocytes from exceptional individuals.[4] Suggestions concerning a chromosomal abnormality as the cause of Down's syndrome (mongolism) were also put forward in the nineteen-thirties, but their validity was not proven for a quarter of a century, until after the correct chromosomal number of man had been discovered.

Biochemical and Molecular Genetics. These areas of human genetics again are too recent to have more than a slight direct relation to Mendel (Fig. 4, right). Nevertheless, on its rediscovery, Mendel's work was of immediate significance to Garrod's studies of errors of metabolism in man (1902). That these errors have a tendency to occur in more than one member of a family was apparent to Garrod in 1899 in the case of alkaptonuria, but its specific Mendelian nature was only recognized three years later. Biochemical genetics, which began with these studies on man, developed in its modern form in studies of flies, the meal moth, the silkworm, and particularly the mold Neurospora (Beadle and Tatum 1941). Stimulated by its concepts, human genetics reentered the field and, with its analyses of inherited variations of the hemoglobin molecules, made pioneering contributions to general molecular biology. Mendel could hardly have foreseen these developments, but there is an inkling of a premonition in his reference to the cell elements in the basic cells which stand in dynamic interaction to one another ("welche in den Grundzellen ... in lebendiger Wechselwirkung stehen").

4. For further references see C. Stern and G. L. Walls, "The Cunier Pedigree of 'Color Blindness,'" *Am. J. Human Genet.* 9(1957): 249–273.

Mendel's Accomplishment. It is possible to regard the growth of knowledge as a superpersonal accomplishment of the Human Mind. In such a view there is little place for heroes and hero-worship. Each investigator contributes only what some other one would also have been able to contribute. And if a specific personality has given a special touch to the discovery, such a feature is nothing but an ephemeral phenomenon. From the point of view of eternity this may well be true. Yet the Human Mind exists only in individuals. The living who search for knowledge can gain guidance and inspiration from their predecessors. Mendel's accomplishment is unique not only for its pioneering success but also for the way in which it was attained. With hardly any prior experience in original studies, for eight years Mendel systematically bred and crossed his plants. Without prematurely talking or writing about the facts he discovered, he thought about them—we do not know for how long. When his search had led him to see the principles which stand behind his observations, when he had made the synthesis between the many bare facts and the few generalities, when he, the physics teacher at the local high school, had found their formulation by means of simple theorems of chance combinations, he wrote a single paper of forty-eight neatly handwritten pages. It is indeed justified that today, after a century, we look up to his example as an ideal of scientific accomplishment.

This evaluation is not changed by the fact that the very qualities which make Mendel admirable as a person were at least in part responsible for the neglect of his discoveries. Had he not been content with writing only a single paper on his experiments with peas, had he instead propounded his theory again and again, the history of genetics might not exhibit its void between 1865 and 1900.

REFERENCES

Adams, J.
 1814 *A Treatise on the Supposed Hereditary Properties of Diseases Based on Clinical Observation.* London: J. Callow.
Bateson, B.
 1928 *William Bateson, F. R. S., Naturalist: His Essays and Addresses together with a Short Account of his Life.* Cambridge, England: Univ. Press.

Bateson, W.
1902 Footnote pp. 133-134 in W. Bateson and E. R. Saunders, Experimental Studies in the Physiology of Heredity. *Reports to the Evolution Committee of the Royal Society, Report I,* Part III: 125-160. Reprinted in R. C. Punnett, ed., *Scientific Papers of William Bateson,* vol. 2, p. 39. Cambridge, England: Univ. Press, 1928.
1909 *Mendel's Principles of Heredity.* Cambridge, England: Univ. Press, p. 226.

Beadle, G. W., and E. L. Tatum
1941 Genetic control of biochemical reactions in Neurospora. *Proc. Nat. Acad. Sci. U.S.* 27: 499-506.

Bell, A. G.
1883 Upon the formation of a deaf variety of the human race. *Mem. Nat. Acad. Sci.* 2: 177-262.

Bemiss, S. M.
1858 Report of influence of marriages of consanguinity upon offspring. *Trans. Amer. Med. Assn.* 11: 319-425.

Bernstein, F.
1924 Ergebnisse einer biostatistischen zusammenfassenden Betrachtung über die erblichen Blutstrukturen des Menschen. *Klin. Wochschr. 3:* 1495-1497.
1925 Zusammenfassende Betrachtungen über die erblichen Blutstrukturen des Menschen. *Z. Induktive Abstammungs-Vererbungslehre 37:* 237-270.

Burks, B. S.
1928 The relative influence of nature and nurture upon mental development: a comparative study of foster parent-foster child resemblance and true parent-true child resemblance. In *27th Yearbook of the National Society for the Study of Education. Nature and Nurture. Part I: Their Influence upon Intelligence* (Bloomington, Ill.: Public School Publ. Co.), pp. 219-316.

Castle, W. E.
1903 The laws of heredity of Galton and Mendel, and some laws governing race improvement by selection. *Proc. Am. Acad. Arts & Sci. 39:* 223-242.
1922 The Y-chromosome type of sex-linked inheritance in man. *Science 55:* 703-704.

Cockayne, E. A.
1933 *Inherited Abnormalities of the Skin and its Appendages.* London: Oxford Univ. Press.

Dahlberg, G.
1929 Inbreeding in man. *Genetics 14:* 421-454.

Danforth, C. H.
1923   The frequency of mutation and the incidence of hereditary traits in man. *Eugenics Genet. Family 1:* 120–128.

Davenport, C. B.
1913   Heredity of skin color in Negro-White crosses. *Carnegie Inst. Wash. Publ. 188.*

Davenport, G. C., and C. B. Davenport
1907   Heredity and eye-color in man. *Science 26:* 589–592.
1910   Heredity of skin pigmentation in man. *Am. Naturalist 44:* 641–672.

Dungern, E. v., and L. Hirschfeld
1910   Über Vererbung gruppenspezifischer Strukturen des Blutes. II. *Z. Immunitaetsforsch. 6:* 284–292.

Dunn, L. C.
1962   Cross currents in the history of human genetics. *Am. J. Human Genet. 14:* 1–13.

Enriques, P.
1922   Hologynic heredity. *Genetics 7:* 583–589.

Farabee, W. C.
1903a Notes on Negro albinism. *Science 27:* 75; W. E. Castle, Note on Mr. Farabee's observations, *ibid.:* 75–76.
1903b Hereditary and sexual influences in meristic variation: A study of digital malformations in man. Ph.D. diss., Harvard University. The genetically relevant part of the thesis appeared in 1905 (see following reference).
1905   Inheritance of digital malformations in man. *Papers Peabody Mus. Am. Archaeol. Ethnol. Harvard 3:* 69–77.

Fischer, E.
1913   *Die Rehobother Bastards und das Bastardierungsproblem beim Menschen; anthropologische und ethnographische Studien am Rehobother Bastardvolk in Deutsch-Südwest-Afrika.* Jena: G. Fischer.

Fischer, M.
1932   Zur Geschichte der Bluterkrankheit. *Eugenik 2:* 119–120.

Fisher, R. A.
1918   The correlation between relatives on the supposition of Mendelian inheritance. *Trans. Roy. Soc. Edinburgh 52:* 399–433.
1934   The effects of methods of ascertainment upon the estimation of frequencies. *Ann. Eugenics 6:* 13–25.

Flemming, W.
1898   Über die Chromosomenzahl beim Menschen. *Anat. Anz. 14:* 171–174.

Galton, F.
1897   The average contribution of each several ancestor to the total heritage of the offspring. [sic] *Proc. Roy. Soc. (London) 61:* 401–413.
1908   *Memories of My Life.* London: Methuen & Co., pp. 294–295.

Garrod, A. E.
1899   A contribution to the study of alkaptonuria. *Med.-Chir. Trans. 82:* 367.
1901   About alkaptonuria. *Lancet 2:* 1484–1486.
1902   The incidence of alkaptonuria: A study in chemical individuality. *Lancet 2:* 1616–1620.
1908   The Croonian Lectures on inborn errors of metabolism, Lecture I (*Lancet 2:* 1–7), pp. 5–6.

Glass, B.
1959   Maupertuis, pioneer of genetics and evolution. In *Forerunners of Darwin: 1745–1859,* ed. B. Glass. O. Temkin, and W. L. Straus, Jr. (Baltimore: Johns Hopkins Press), pp. 51–83.

Goddard, H. H.
1914   *Feeble-mindedness, its Causes and Consequences.* New York: Macmillan Co.

Gunther, M., and L. S. Penrose
1935   The genetics of epiloia. *J. Genet. 31:* 413–430.

Haldane, J. B. S.
1932a   A method for investigating recessive characters in man. *J. Genet. 25:* 251–255.
1932b   Genetical evidence for a cytological abnormality in man. *J. Genet. 26:* 341–344.
1935   The rate of spontaneous mutation of a human gene. *J. Genet. 31:* 317–326.
1938   *Heredity and Politics.* New York: Norton.

Haller, M. H.
1963   *Eugenics: Hereditarian Attitudes in American Thought.* New Brunswick, N.J.: Rutgers Univ. Press.

Hardy, G. H.
1908   Mendelian proportions in a mixed population. *Science 28:* 48–50.

Hirschfeld, L., and H. Hirschfeld
1919   Serological difference between the blood of different races. *Lancet 2:* 675–679.

Hogben, L.
1931   The genetic analysis of familial traits. I. Single gene substitutions. *J. Genet. 25:* 97–112.
1933   *Nature and Nurture.* New York: Norton.

Horner, J. F.
1876   Ein Beitrag zum Vererbungsgesetz. Bericht über die Verwaltung des Medizinwesens des Kanton Zürich vom Jahre 1876.

Huntington, G.
1872   On chorea. Med. Surg. Reptr. 26: 317–321.

Hurst, C. C.
1908   On the inheritance of eye colour in man. Proc. Roy. Soc. (London), Ser. B 80: 85–96.

Iltis, H.
1924   Gregor Johann Mendel: Leben, Werk und Wirkung. Berlin: J. Springer, p. 151.

Landsteiner, K.
1900   Zur Kenntnis der antifermentativen, lytischen und agglutinierenden Wirkungen des Blutserums und der Lymphe. Zentr. Bakteriol. 27: 357–362.
1901   Über Agglutinationserscheinungen normalen menschlichen Blutes. Wien. Klin. Wochschr. 14: 1132–1134.

Lawrence, E. M.
1931   An investigation into the relation between intelligence and inheritance. Brit. J. Psychol. Monogr. Suppl. No. 16, pp. 1–80.

Lenz, F.
1919   Die Bedeutung der statistisch ermittelten Belastung mit Blutsverwandschaft der Eltern. Münch. Med. Wochschr. 66, 2: 1340–1342.

Levit. S. G.
1936   The problem of dominance in man. J. Genet. 33: 411–434.

Maupertuis, P. L. M. de
1745   Vénus physique, contenant deux dissertations, l'une sur l'origine des hommes et des animaux; et l'autre sur l'origine des noirs. The Hague. Pt. I, chap. 17.

Mendel, G. Letters of 3 July and 27 September 1870 in C. Correns (ed.), Gregor Mendel's Briefe an Carl Nägeli, 1866–1873 (Abhandl. Math.-Phys. Kl. Königl. Sächsischen Ges. Wiss. 29 [1905]: 189–265), pp. 229–242. English translation by L. K. Piternick and G. Piternick in: The Birth of Genetics, Genetics 35 (1950, suppl.): 1–29; also in The Origin of Genetics, ed. C. Stern and E. R. Sherwood (San Francisco: W. H. Freeman, 1966).

Morgan, T. H.
1911   The application of the conception of pure lines to sex-limited inheritance and to sexual dimorphism. Am. Naturalist 45: 65–89.

Muller, H. J.
1925   Mental traits and heredity. J. Heredity 16: 433–448.
1927   Artificial transmutation of the gene. Science 66: 84–87.

Nasse, C. F.
1820　Von einer erblichen Neigung zu tödtlichen Blutungen. *Arch. med. Erfahr. 1820:* 385–434.

Newman, H. H., F. N. Freeman, and K. J. Holzinger
1937　*Twins: A Study of Heredity and Environment.* Chicago: Univ. of Chicago Press.

Painter, T. S.
1923　Studies in mammalian spermatogenesis. II. The spermatogenesis of man. *J. Exptl. Zool. 37:* 291–338.

Pearson, K.
1904　Mathematical contributions to the theory of evolution. XII. On a generalised theory of alternative inheritance with special reference to Mendel's laws. *Phil. Trans. Roy. Soc. (London), Ser. A 203:* 53–86.

(ed.) 1912–　*The Treasury of Human Inheritance.* Eugenics Laboratory Memoirs, Francis Galton Laboratory for National Eugenics, University of London.

Pearson, K., E. Nettleship, and C. H. Usher
1911–1913　*A Monograph on Albinism in Man.* Drapers' Company Research Memoirs, Biometric Series 6, 8, 9. London: Dulau and Co.

Punnett, R. C.
1908　Mendelism in relation to disease. *Proc. Roy. Soc. Med. 1 Epidemiol. Sect.:* 135–161, followed by discussion, *ibid.:* 161–168.

1915　*Mimicry in Butterflies.* Cambridge, England: Univ. Press, App. I, pp. 154–156.

1917　Eliminating feeblemindedness. *J. Heredity 8:* 464–465.

1950　Early days of genetics. *Heredity 4:* 1–10.

Réaumur, R. A. F. de
1751　*L'Art de faire eclorre et d'élever en toute saison des oiseaux domestiques de toutes espèces,* 2 ed., 2 vols. Paris: L'Imprimerie Royale, p. 2.

Sedgwick, W.
1861　On sexual limitation in hereditary disease. *Brit. Foreign Med.-Chir. Rev.* (American reprinting) *27:* 245–253; *28:* 140–151. As cited by R. R. Gates, 1923, in *Heredity and Eugenics* (London: Constable) *27:* 477–489; *28:* 198–214.

Siemens, H. W.
1924　*Die Zwillingspathologie.* Berlin: J. Springer.

Stern, C.
1943　The Hardy-Weinberg law. *Science 97:* 137–138.

Tjio, J. H., and A. Levan
1956　The chromosome number of man. *Hereditas 42:* 1–6.

Verschuer, O. v.
1939 Twin research from the time of Francis Galton to the present day. *Proc. Roy. Soc. (London), Ser. B 128:* 62–81.

Weinberg, W.
1908 Über den Nachweis der Vererbung beim Menschen. *Jahresh. Ver. vaterl. Naturk. Württemb. 63:* 369–382.
1909 Über Vererbungsgesetze beim Menschen. *Z. Induktive Abstammungs-Vererbungslehre 1:* 377–392, 440–460; *2:* 276–330.
1910 Weitere Beiträge zur Theorie der Vererbung. *Arch. Rass. Ges. Biol. 7:* 35–49, 169–173.
1912a Über Methode und Fehlerquellen der Untersuchung auf Mendelsche Zahlen beim Menschen. *Arch. Rass. Ges. Biol. 9:* 165–174.
1912b Zur Vererbung des Zwergwuchses. *Arch. Rass. Ges. Biol. 9:* 710–717.

Weinstein, A.
1933 Palamedes. *Am. Naturalist 67:* 222–253.

Whisson
1778 An account of a remarkable imperfection of sight. In a letter from J. Scott to the Rev. Mr. Whisson, of Trinity College Cambridge. Communicated by the Rev. Michael Lort, B.D., F.R.S. *Phil. Trans. Roy. Soc. (London) 68:* 611–614.

Wilson, E. B.
1911 The sex chromosomes. *Arch. mikroskop. Anat. 77, Abtl. 2:* 249–271.

Winiwarter, H. v.
1912 Études sur la spermatogenèse humaine. *Arch. Biol. (Liège) 27:* 91–189.

Yule, G. U.
1902 Mendel's laws and their probable relations to intra-racial heredity. *New Phytologist 1:* 193–207, 222–238.

Genetic Mosaics in Animals and Man

A LANSQUENET BEARS A CHILD

A weird happening has occurred in the case of a lansquenet [soldier] named Daniel Burghammer . . . When the same was on the point of going to bed one night he complained to his wife, to whom he had been married by the Church seven years ago, that he had great pains in his belly and felt something stirring therein. An hour thereafter he gave birth to a child, a girl . . . He then confessed on the spot that he was half man and half woman . . . He also stated that . . . he only slept once with a Spaniard, and he became pregnant therefrom. This, however, he kept a secret unto himself and also from his wife, with whom he had for seven years lived in wedlock, but he had never been able to get her with child . . . The aforesaid soldier is able to suckle the child with his right breast only and not at all on the left side, where he is a man. He has also the natural organs of a man for passing water . . . All this has been set down and described by notaries. It is considered in Italy to be a great miracle and is to be recorded in the chronicles. The couple, however, are to be divorced by the clergy.—From Piadena in Italy, the 26th day of May 1601.

—*The Fugger News-Letters,* ed. Victor von Klarwill

Multicellular plants and animals are compounds of different organs, tissues, and cells. This compoundedness is an orderly one resulting from the normal processes of development: an activated egg cell divides and the products of its division multiply further and differentiate. In these divisions the nucleus of the activated egg cell replicates by mitosis so that the genetic endowment of the initial nucleus is transmitted to all descendent nuclei. The orderly processes of differentiation thus are

27

superimposed on the genetic identity of the compounding cells.

There are exceptions, still orderly, to this usual situation. In some few animals specific chromosomes are extruded from certain body cells during development, or specific chromosomes are selectively multiplied. In this way regular differences between somatic nuclei and nuclei in cells of the germ line become established. Moreover, even when the initial chromosomal endowment of the fertilized egg cell is replicated normally and transmitted to all cells of the resulting individual, there is need to scrutinize the statement that differentiation is superimposed on the genetic identity of the compounding cells. This identity is potential only in the sense that the coded information in the DNA of the chromosomes presumably is the same in all cells. Actually, however, differentiation seems to come about through evocation or inhibition of specific genic potentialities out of the totality of potentialities. Thus, differentiation depends on and results in differences in the effective genic constitution of different cells.

Aspects of the genetics of the normal, regular compoundedness of differentiated organisms will be the subject of the third of these essays. In the present essay we will survey those curious creatures which have long aroused the interest of biologists and people in general because they are irregular compounds, seemingly created "against Nature." Some of them are individuals composed of male and female parts that normally characterize one or the other sex only. Others are compounded of two or more different, but not sex-related, genetic constitutions, each of which would usually be found in a separate individual. Our path will lead beyond such exceptional organisms. We will see that certain types of mosaicism occur as normal phenomena in whole groups of animals including man.

A Few Examples of Mosaics. Since ancient times artists have given expression to fantasies in which male and female traits were united in imaginary human hermaphrodites. Greek sculptors attempted to depict harmonious bodies with attributes of both sexes (Fig. 5), and Hindu artists created statues of Ardhanarisvara, "half woman, half god," in which the deity

Shiva and his wife, Parvati, were shown "in one": one side of the body male, the other female (Fig. 6).

Human hermaphrodites exist indeed as living individuals, but the reality is far removed from harmony. Male and female body parts occur in irregular compositions and incongruous associations. One individual may have incomplete female ducts but male gonads, another mostly male ducts but a testis on only one side, an ovary on the other. A third may possess one normal ovary or testis plus a compound oötestis, a fourth may have two oötestes.

One of the earliest descriptions of an abnormal individual, part male and part female, was that published in 1761 by Jacob Christian Schäffer in Germany under the title, *Der wunderbare und vielleicht in der Natur noch nie erschienene Eulenzwitter* (The miraculous moth hermaphrodite which perhaps never before had appeared in Nature). This was a specimen of *Lymantria dispar,* the gypsy moth. The normal sexes of this moth differ greatly in the form of the antennae and the size and coloration of the wings. The miraculous specimen had on its right side an antenna and wing typical for "a true and honest male" and on its left side those typical for "a true and honest female." "How marvelous," wrote Schäffer, "is not Nature in all her works and products! Not only does she create objects according to the orderliness and the rules established by herself

FIG. 5. Hermaphrodite. Outline drawing of an antique marble statue. (From F. de Clarac 1836–1837, no. 1546 D.)

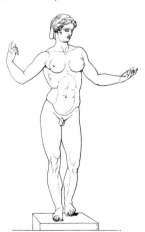

. . . She even knows how to combine and join together objects in such ways that human intelligence must stop in its path and must exclaim with astonishment: What a wonder of Nature! How did this happen? Is there a purpose in it and an intention? How? Did this originate without reason and just by chance—or according to rules in a natural way and according to its own determination?"

The future would bring even more striking examples of sexual and nonsexual mosaics, a term which denotes compounds consisting of two (or, rarely, more) genetically different

FIG. 6. Ardhanarisvara. Hindu statue (early Khmer art), 6–8 century A.D. Note the moustache and male torso at left, and the absence of moustache and presence of a female breast at right. The garment around the waist at left signifies the tiger skin worn by the male god Shiva, at right shows decorations characteristic of a female. (From the collection of the National Museum, Bangkok, Thailand.)

tissues. Sex mosaics, called gynandromorphs or gynanders (from the Greek *gyne*, female, and *aner*, *andros*, male), are particularly impressive in species whose normal sexes differ greatly in their secondary sex characteristics. Thus, the male stag beetle has large antler-like mandibles and broadly laminated antennae, while the mandibles and antennae of the female are small. A gynandric specimen may have female appendages on one side and male appendages on the other (Fig. 7). Or, the female of a species of solitary wasps is colored bright red and lacks wings, while the male is black and winged. A gynander has been found whose left half, red and wingless, is joined to a black and winged right half (Fig. 8). In one species of butterfly, the hind wing of the male has a swallow tail, and that of the female is of more usual shape. Moreover, most of the upper surface of the male's wings is colored light yellowish while the wings of the female are mostly black. A gynander may seem to be artificially made up of two separate left and right halves, one derived from each sex (Fig. 9A). In another species, a gynander may show the white wings of the female on one side and the orange-colored ones of the male on the other (Fig. 9B).

While half-and-half sex mosaics are particularly impressive, they are by no means the only ones. Whenever an apparently fully bilateral gynander is scrutinized carefully, some body region will be found where one of the two sexually different

FIG. 7. Head and prothorax of a gynandric stag beetle *Lucanus cervus*. Right: male, head with sculptured ridges, large mandible, long antenna; left: female, head without ridges, small mandible, and short antenna. (From Dudich 1923.)

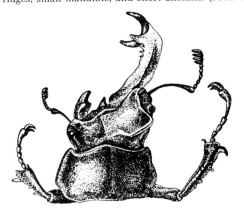

parts extends beyond the midline. Moreover, many gynanders may deviate from bilaterality to a great degree. In Drosophila, for instance, the whole head and thorax may be male, the abdomen female, or the reverse arrangement may occur. In addition, the relative amounts of male and female tissue may vary a great deal, some gynanders being mainly female and others mainly male. Finally, it should be added that not all mosaics are compounded of two or a few coherent masses of genetically different tissues. Sometimes intricate mixtures of these tissues occur, as on the wings of the gynandric butterfly shown in Figure 9C.

Gynanders in the Honeybee. The origin of gynanders had to remain obscure until the mechanisms and the physiology of normal sex determination were known. These were first discovered in the honeybee. In the early years of the nineteenth century a German schoolmaster, Johann Gottfried Lukas, reported that his beehives had produced a number of "sting drones," animals which combined traits of the normally stingless males with those of the stinged female worker bees. He gave careful descriptions of the mosaic appearance of his strange specimens. Some had all male legs on one side and all worker legs on the other; others had six worker's legs but two great drone eyes; still another had a drone eye on the left and a worker's eye on the right side of the head. Furthermore, Lukas discovered that the mosaics came from the small cells of the bee combs,

FIG. 8. Gynander of the solitary wasp *Pseudomethora canadensis*. Right: male, black coloration, 13 antennal segments, 2 ocelli, wings, male-type legs. Left: female, bright red coloration, 12 antennal segments, no ocelli, no wings, female-type legs. (Redrawn from Wheeler 1910).

which ordinarily house the larvae and pupae destined to develop into workers, rather than the larger cells in which drones develop. In spite of his careful observations his work was attacked as being "a lie and pitiable blabber," and for more than fifty years no one either found or dared to describe new sting drones.

In 1860, the German beekeeper Johann Jacob Eugster discovered new gynandric bees in one of his stocks. This time they did not fail to gain the attention of the scientific world. Theodor von Siebold, the famous professor of zoology at the University of Munich, gave a detailed description of Eugster's hermaphroditic specimens (1864) and came to the defense—posthumously—of the maligned Lukas. One of von Siebold's fields of studies was parthenogenesis. In the analysis of this phenomenon, the determination of sex in honeybees had been an exciting and much-discussed topic. Another beekeeper, the Lutheran minister Johann Dzierzon, who had revolutionized apiculture by his treatises on a new kind of beehive as well as by his biological insights, had proposed that queen and worker bees come from fertilized eggs but that drones arise

FIG. 9. Three gynandric butterflies. (A) *Papilio dardanus*. Left, female; right, male. (B) *Colias eurytheme*. Left, female with white background; right, male with orange background. (C) *Gonepterix cleopatra*. Irregular mosaicism of male parts (orange and yellow, stippled) and female parts (near white, not stippled). (Redrawn from A. Hannah-Alava, "Genetic Mosaics." Copyright © 1960 by Scientific American, Inc. All rights reserved.)

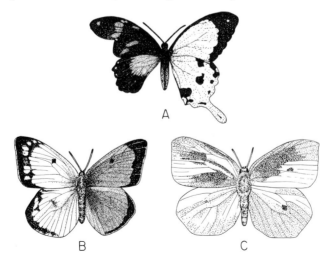

parthenogenetically. This theory, the first valid explanation of sex determination in any organism, led von Siebold to an attempt to explain the origin of bee gynanders. He believed that in most animals a minimum mass of sperm is necessary for fertilization. Too small an amount usually would not result in development. In the bee, however, an egg is able to develop without sperm, such development resulting in "only a male." Fertilization with full quantity of sperm brings forth reversal of the male tendency to the female tendency. It seemed to him, therefore, that fertilization with insufficient amount of sperm would only partially overcome the tendency toward male development so that, instead of a normal female resulting from such fertilization, a hermaphroditic mixture of male and female parts would be formed.

It is no wonder that von Siebold did not find a satisfactory solution to his problem. Not until 1875 was it shown that fertilization of an egg consists in the fusion of a single nucleus derived from a single sperm with the nucleus of the egg itself. This was discovered in the gametes of sea urchins. Thirteen years later Boveri observed some abnormal embryos of these animals which led him to an explanation of bee gynanders (1888). Sometimes, he found, the sperm nucleus does not immediately fuse with the egg nucleus. Instead, the latter divides by itself and the resulting two blastomeres differ in that one contains one of the division products of the egg nucleus only, while the other contains both the other division product and the sperm nucleus. This nucleus now fuses with the derived egg nucleus in its blastomere and the resulting embryo becomes a mosaic of cells with unfertilized haploid nuclei and cells with fertilized diploid nuclei (Fig. 10). If such "delayed fertilization" occurred in a honeybee egg, the cells with unfertilized nuclei would give rise to male parts and those with fertilized nuclei to female parts (Boveri 1889).

Boveri realized that his explanation could be tested if the queen bee whose offspring included the gynanders had been mated to a drone who carried a different genetic marker than she herself. The male parts of the gynanders should exhibit only the phenotype of genes derived from the mother while the female parts should be hybrid in character. Eugster's hermaphrodite bees had indeed been derived from a cross between a lightly colored Italian queen and a darker colored German drone, but for a long time Boveri was unable to locate the gynanders. More than twenty years later, close to the end of his life, the bottle was found which contained Eugster's preserved specimens. An analysis of their coloration convinced Boveri (1915) that indeed the male parts were purely maternal in color and the female parts of hybrid type. Boveri was a meticulous worker, but it is difficult to decide whether his conclusion was fully justified since time had dealt badly with these gynanders.

More recently, strains of bees have been found which are genetically predisposed to form many sex mosaics. In addition, bee gynanders can be obtained experimentally by chilling the eggs (Drescher and Rothenbuhler 1963). Among 82 gynanders obtained after such treatment, 80 had male parts of paternal and female parts of hybrid phenotype, the opposite of Boveri's expectation. An explanation appropriate to the new findings and first proposed on general grounds by T. H. Morgan (1905) makes use of the fact that polyspermy occurs regularly in inseminated bee eggs. Normally only a single sperm participates in development, the supernumerary ones undergoing degeneration. If, however, one of the extra sperm nuclei should divide and continue to do so, its descendants would build up haploid male tissue while the descendants of the fertilized egg nucleus would form diploid, female tissue (Fig. 11).

The two exceptional gynanders of Drescher and Rothenbuhler corresponded to Boveri's expectation in that the male

FIG. 10. A possible mode of origin of a gynandric bee. The egg nucleus divides before fertilization and a sperm nucleus fuses with one of the two egg nuclei. The resultant gynander is hybrid in its female and purely maternal in its male parts.

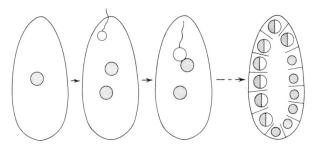

parts were maternal in phenotype. Their origin may indeed have been a consequence of delayed fertilization, or perhaps of a slightly different process, namely normal fertilization of the egg nucleus as well as development of a polar nucleus. Still other variations on the theme of abnormal egg and sperm behavior in bee development are possible. For example, using mixed sperm from two genetically different males in artificial insemination, mosaics could be recognized in which three separate sperm nuclei participated in development (Rothenbuhler 1958). In these experiments the queen parent was homozygous for the recessive ivory eye coloration and the mixed sperm came from ivory and chartreuse drones. The gynanders had ivory or black female eye tissue depending on whether fusion with the egg nucleus had involved an ivory- or a chartreuse-carrying sperm. The male parts, in some mosaics, were themselves a mixture of ivory and chartreuse tissues. Apparently one accessory sperm may give rise to the ivory and another to the chartreuse male tissue of the triple mosaic. A still less conventional type of fertilization is illustrated by some gynanders whose male parts were maternal and whose female parts must have been derived from nuclei that were formed by the fusion of two different sperm nuclei within the egg (Laidlaw and Tucker 1964). Parts of these organisms had a female parent only and other parts two male and no female parent!

As we have seen, there are a variety of ways by which gynandric bees can arise. We may cite Cooper (1959) for a

---

FIG. 11. A possible mode of origin of a gynandric bee. This is an alternative to the mode illustrated in Fig. 10. Two sperm nuclei participate in development. One fuses with the egg nucleus; the other is replicated without fusion. The resultant gynander is hybrid in its female and purely paternal in its male parts.

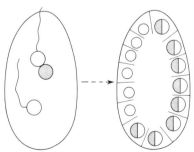

succinct summary of the cytologic origins of gynandric bees and gynandric hymenoptera in general. They "may arise in a number of different ways, all of which involve either oögenetic meiotic mishaps, failures to suppress the activity of all but one product of these meioses, or failures to arrest the activity of sperm not participating in syngamy, or combinations of these events."

Further Mosaics from Supernumerary Nuclei. Occasional participation in development of supernumerary sperm is not restricted to the honeybee. Evidence for its occurrence comes from various other insects and even from a vertebrate. In the pigeon, W. F. Hollander (1949) has described a number of mosaics which can best be interpreted by assuming that not only typical fusion of an egg and a sperm nucleus had occurred but that in addition one or more sperm nuclei independently took part in development. It is unknown whether these nuclei and their descendants remained haploid or whether only those nuclei survived that happened to become diploid.

Most of the haploid-diploid mosaics of *Drosophila pseudoobscura* described by Crew and Lamy (1938) seem to have originated through a variation on the theme of development of tissues from a sperm nucleus. In these specimens the diploid nuclei carried the same paternal genome as the haploid nuclei, although on the basis of random selection of a second sperm a different genome would have been expected in half of the mosaics. Crew and Lamy assume that in these cases the male pronucleus (from a single sperm) divided before having fused with the female pronucleus. The latter either failed to divide at this stage and was subsequently fertilized by one of the division products of the sperm, or, if it did divide, only one of its products was fertilized and the other degenerated.

Sex determination in the honeybee is not the simple haploidy-diploidy mechanism that Dzierzon's theory implied. Rather, maleness and femaleness depend on a peculiar genetic alternative which was first brilliantly elucidated by P. W. Whiting in the parasitic wasp Habrobracon. Whiting had found the usual diploid females and haploid males and, in addition, in some of his cultures biparental, diploid males. An explanation of the origin of these unexpected individuals was first suggested to him by some strange sex mosaics that were named gynandroids (Whiting, Greb, and Speicher 1934). These animals were mostly male but the maleness itself was compounded of two different genotypes. If, in respect to a certain gene or gene complex X, the diploid female parent was heterozygous, for example $X^a X^b$, then the gynandroid mosaics were haploid X^a in one part of their body and haploid X^b in the other. Apparently they originated parthenogenetically from an egg with two different functional nuclei that had been the result of meiosis. The remarkable fact which gave the gynandroids their name was the appearance of female-type structures at the meeting line of the two different male tissues (Fig. 12). From this fact Whiting (1940) reasoned that femaleness was the result of complementation by the two different genetic determinants X^a and X^b. In a normal female this complementation occurs within each $X^a X^b$ cell, while in a gynandroid it is dependent on diffusion or transport of complementary gene products from X^a to X^b cells and vice versa. The diploid males found by Whiting are homozygous $X^a X^a$ or $X^b X^b$ (or $X^c X^c$ . . . $X^n X^n$). All such indi-

FIG. 12. *Habrobracon juglandis.* (A) Ventral view of the genitalia of a gynander. A complete set of male genitalia is present. The left sensory gonapophysis (g) of normal length and elements of the sting (s) are the female structures. (B) Ventral view of the genitalia of a gynandroid, a mosaic composed of two different haploid male sex genotypes. A complete set of male genitalia is present, but the right outer clasper is feminized and resembles a minute sensory female gonapophysis (g). (Redrawn from Figs. 1 and 2, Whiting, *J. Morphol.* 66: 326.)

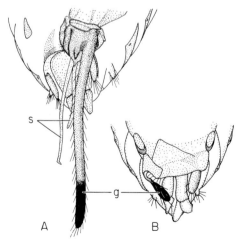

viduals lack a complementing genotype as do the usual haploid X^a and X^b (and X^c . . . X^n) males. It was later shown that in the honeybee sex is determined in a way very similar to that in Habrobracon.

Gynanders Produced by Chromosome Elimination. In most animals parthenogenesis is not involved in the normal decision for maleness or femaleness, and thus the explanation of the origin of gynanders had to be sought in other mechanisms. Time was not ready for the solution of this problem until the first decade of the present century, when sex chromosomes were discovered and the difference between the sexes in most animals was traced back to the presence or absence, or the presence of varying numbers, of such chromosomes. The relation was obvious in grasshoppers and those other forms of animals in which one sex has two and the other only one of the X-chromosomes, without the complicating fact of the presence of a Y-chromosome in one of the two sexes. Here clearly the balance of the two sexual tendencies in development is decided by the number of X-chromosomes. In contrast, it is not obvious how sex is determined when one sex is XX and the other XY. Is a Drosophila female a female because she has two X-chromosomes instead of one, or because she lacks a Y-chromosome? Is a Drosophila male a male because he has only one X-chromosome or because he has a Y? Or do both X and Y interact to make 2X = ♀ and XY = ♂? Only exceptional individuals with chromosome constitutions such as XXY, XXYY, and XO (where O stands for absence of any additional sex chromosome) permitted an answer to these questions. In Drosophila, it was found, the number of X-chromosomes is the decisive element independent of the Y-chromosome, because XXY and XXYY flies are females and XO flies males. The Y-chromosome, while necessary for complete spermatogenesis, is not needed for male morphology and behavior. In man the XX–XY alternative works differently. Apart from the normal XX women, the unusual XO and XXX types are females, and apart from the normal XY men, the unusual XXY, XXYY, XXXY, and other Y-carrying types are males. It is the presence or absence of the Y-chromosome that leads to one or the other sex.

These insights into the at least superficially simple role which the quantity or the presence of specific chromosomes plays in the differentiation of sexual traits became the foundation for new interpretations of gynandrism. In those organisms in which parthenogenesis cannot be invoked, gynanders are the result of the presence of two different sex chromosomal constitutions in the body of a single individual. Thus, in Drosophila, one part of the individual gynander may consist of XX cells, another of XO or XY cells. The part with XX cells will give rise to female structures, that with XO or XY cells to male structures. Direct proof by chromosome count and inspection is difficult to obtain in adult flies, because cell divisions in somatic tissues no longer take place. Even the gonads of gynanders usually are not favorable for chromosomal study. Where cytology fails, however, genetics can furnish equivalent evidence. Genes are markers of chromosomes and, specifically, X-linked genes are markers of X-chromosomes. If the parents of a gynander carry suitable X-linked genes, its female and male parts may not only be recognizable by their sexual features but also by those traits which are determined by the genic markers. One might take, for example, the cross of a Drosophila female which is homozygous for the two recessive X-linked genes determining eosin eyes and miniature wings to a male which carries in his single

FIG. 13. Gynander of *Drosophila melanogaster*. Its right side is male with light eosin eye color, miniature wing, and sex comb on foreleg; its left, female with red eye color, long wing, and absence of sex comb. The maternal X-chromosome carried the dominant alleles for not-eosin and not-miniature, the paternal eosin and miniature. The gynandric condition arose by elimination of the maternal X-chromosome. (Redrawn from Morgan and Bridges 1919.)

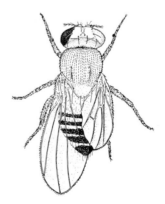

X-chromosome the dominant alleles leading to red eyes and long wings. The daughters of these flies will carry one X-chromosome with eosin eyes and miniature wings and another with red eyes and long wings. Their phenotype is the dominant: red and long. Their brothers obtain their single X-chromosome from the mother. They have eosin eyes and miniature wings. The rare gynanders which may be encountered in a large experiment are of two kinds. Both are red, and long in their relevant female sections. About one-half of them will be of the same appearance in their relevant male parts, the other half will be eosin and miniature wherever maleness exists (Fig. 13). Apparently, the relevant female parts are heterozygous for the two genes, as is appropriate for whole females from the cross, but the male parts are either hemizygous for the X-chromosome derived from the father or for that derived from the mother. Morgan and Bridges, who analyzed many Drosophila gynanders, came to the conclusion that the great majority of them begin life endowed with two X-chromosomes (1919). Then, usually during the first division of the fertilized egg, only one of the X-chromosomes divides normally (Fig. 14). Its division

FIG. 14. Elimination of an X-chromosome during the first mitosis of a developing egg of *Drosophila melanogaster*. (A) Loss of maternal chromosome (stippled). (B) Loss of paternal chromosome (solid). The size of the mitotic figures relative to the whole egg is greatly exaggerated in these diagrams.

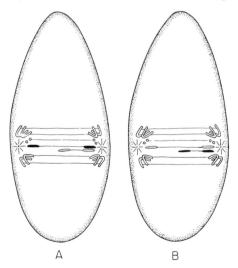

A   B

products move to opposite poles of the mitotic spindle so that each daughter nucleus receives one of the daughter chromosomes. The division products of the other X-chromosome, however, do not both succeed in being included in a nucleus. One of them does so but the other lags behind and degenerates in the cytoplasm. In this way two nuclei of different genotypes are established within the egg, one with two X-chromosomes and one with a single X. Their nuclear descendants lead to a gynander with the phenotype of the male parts depending upon whether the maternal or the paternal X-chromosome was eliminated.

The non-inclusion of an X-chromosome into its appropriate cleavage nucleus is not always due to its lagging behind on the spindle and its ultimate degeneration in the cytoplasm. Instead, in a few instances it has been possible to show by means of genetic markers that two sister X-chromosomes come to be included in the same nucleus together with one of the other, normally distributed sister X-chromosomes. The result is a triple mosaic consisting of XX tissue derived from nuclei that never underwent abnormal chromosome distribution, XO tissue derived from a nucleus that did not receive either one of the sister chromosomes of one of the X-chromosomes, and XXX tissue derived from the nucleus that received both of these sister chromosomes (Stern 1960b).

Morgan and Bridges' theory of gynandrism in Drosophila implies that the male parts of the gynanders lack a Y-chromosome. They therefore should be sterile as males even if the distribution of the male tissues permitted formation of morphologically normal male genitalia and testes. Indeed, most gynanders are sterile as males, but are fertile as females if they have normal female sex organs. Only very rarely have gynanders produced offspring as males from normal testes. In such cases there was reason to believe that the initial nucleus of the fertilized egg had the constitution XXY so that the elimination of an X-chromosome led to the sex mosaicism XXY–XY, which in suitable cases permits fertility as either a female or a male, depending on which chromosome constitution is present in the genitalia and gonads.

Behavior of Gynanders. Fertility or sterility of gynanders does not depend only on the presence or absence of a Y-chromosome, nor only on the existence of a morphologically adequate system of external genitalia, internal ducts, and functional gonads. Behavioral aspects are also involved. Mating in Drosophila, for instance, is preceded and accompanied by involved courtship maneuvers in which the roles of the two sexes are quite different. Mating behavior in gynanders may be of all possible types including indifference to either sex, male courtship, female courtship, and courtship as both male and female. There is only fragmentary knowledge on possible correlation between type of courtship and distribution of the male and female areas. In any case these two genotypes can primarily be distinguished on the surface of the animal only and give at best an incomplete picture of internal mosaicism. Nevertheless, an embryological correlation exists between external and internal genotype of mosaics in a given body region. Thus, an externally male head is more likely to enclose a male than a female brain and vice versa. There is indeed some evidence for an influence of the sex of the head, as visible externally, on genital function (L. V. Morgan 1929). Unfortunately, the record does not provide direct observations on mating behavior but only distinguishes fertility and sterility. As Table 1 shows, among 34 gynanders with externally normal female abdomens, 19 were fertile and 15 sterile. Among the fertile individuals, 7 had female heads and only 1 a male head, whereas 2 sterile individuals had female heads and 6 had male heads. In the case of externally mixed heads, fertility and sterility were about equally frequent.

More detailed accounts of the behavior of gynanders have been provided for the wasp *Habrobracon juglandis* (Whiting 1932). They bear out the evidence gained from Drosophila that,

TABLE 1. Reproductive capacities of *Drosophila melanogaster* gynanders with normal female abdomens.

Sex of head	Number of gynanders	
	Fertile	Sterile
Female	7	2
Mixed	11	7
Male	1	6

Source: Data from L. V. Morgan 1929.

in general, the sex of the head, regardless of the sex of the abdomen, determines the type of sex-specific reactions. The behavioral differences in these wasps are not restricted to courtship and mating. Males show mating response toward females, but are indifferent to caterpillars, which are the hosts of the parasitic larvae of Habrobracon. Females, on the other hand, not only respond to the courtship of males, but show complex stinging and egg-laying reactions toward caterpillars. All 29 gynanders with male heads reacted positively to females, and all 24 of these gynanders who were placed with caterpillars were indifferent to them. Only 4 gynanders had female heads. They were indifferent to other females and reacted positively to caterpillars. Among the 17 gynanders with mixed heads some reacted as males and some as females. Observations suggested that sexual behavior is more closely correlated to the sex of the antennal and ocellar regions of the head than to the regions of the compound eye. Particularly interesting are some gynanders with mixed heads that showed abnormal reactions. Whiting writes of them as follows:

A few instances are reported of atypical behavior of gynandromorphs; "non-stop" in which certain individuals were stimulated by females but failed to mount, running back and forth repeatedly; "momentary reversal" in which response characteristic of one sex was shown momentarily by gynandromorph with instincts in general of opposite sex; "bisexual" in which the same individual consistently gave female reaction toward caterpillars and male reaction toward females; "wires crossed" in which mating response was given toward a caterpillar or stinging reaction was shown toward a female. Such irregularities are regarded as due to mosaic character of sensory and nervous system.

Gynanders from Double-Nucleus Fertilization. The theory that gynandrism in Drosophila is due to elimination of one X-chromosome from an initially XX egg was supported by evidence that the autosomes of the egg divided normally and thus remained in their normal diploid state. This evidence was based on information on genetic markers located in the autosomes. It demonstrated that the female and male parts of the gynanders did not differ in their autosomal genotype. Occasionally, however, autosomal mosaics do occur. Some of them are also sex mosaics, others are uniformly female or male. Only a few of these mosaics will be described. Many more, of diverse and sometimes bizarre types, have been treated in L. V. Morgan's monograph on "Composites of *Drosophila melanogaster*" (1929).

It would seem reasonable to assume that autosomal mosaics originate in a way analogous to that of X-chromosomal mosaics, namely by elimination of an autosome. This, however, is not possible, at least for the long autosomes of Drosophila, numbered 2 and 3; independent evidence proves that absence of one chromosome of either pairs is not compatible with survival. It is true that the evidence derives from the fate of fertilized eggs only and not from the fate of parts of mosaic individuals, but there is no reason to doubt that it applies to such parts as well. The imbalance in genic action resulting from a "removal" of one X-chromosome from a 2X female leads to nothing more aberrant than maleness, an evolutionarily satisfactory solution. On the other hand, selective processes have not prepared most species for adjustments to removal of large autosomes. Loss of an autosome may be tolerated, however, when it represents a small part of the whole chromosomal material. This is true particularly for organisms with numerous chromosomes, for which loss of one may have relatively minor consequences. It also applies to Drosophila in the case of the small "fourth" chromosome. Fertilized eggs of Drosophila which carry only a single one of the two "dot" chromosomes normally present are viable and develop into adults, though they are morphologically and physiologically abnormal and very weak. In view of the survival of whole nonmosaic individuals with only one chromosome 4, it is not surprising that mosaics for the phenotype normal (diplo-4) and abnormal (haplo-4) have been observed. One such individual was fertile and produced offspring of both phenotypes. Cytologically the offspring proved to possess both chromosomes (normal) or only one fourth chromosome (abnormal) (Mohr 1932.)

If chromosome elimination of the large autosomes of Drosophila cannot be invoked to explain the origin of mosaics involving these chromosomes, an alternative interpretation is available that circumvents the problems of genic imbalance. This interpretation may be demonstrated in the case of a specimen which was a fertile male with normal pointed bristles on the right half of its thorax and on the whole of its abdomen, and stubble bristles on its head and the left half of its thorax. Its left wing was straight, its right wing curly. The mother of the mosaic had been heterozygous for the dominant gene Stubble in chromosome 3, Sb/Sb^+, and the father had been heterozygous for the dominant gene Curly in chromosome 2, Cy/Cy^+. Apparently the mother had provided two genetically different, fertilizable egg nuclei, one with Sb and one with Sb^+, and the father had provided two genetically different fertilizing sperm, one with Cy, the other with Cy^+ (Fig. 15). Fertilization had resulted in two diploid nuclei, one containing Cy^+ and Sb and the other Cy and Sb^+. Obviously both sperm carried a Y-chromosome because the mosaic was male throughout. Had both carried an X-chromosome, a female would have developed; and had one been an X-sperm and the other a Y-sperm, the animal would have been a gynander as well as an autosomal mosaic.

In the silkworm, *Bombyx mori*, Goldschmidt and Katsuki (1928, 1931) discovered a strain in which mosaicism was caused by a single recessive gene when homozygous in the mother. Offspring of such a parent were either nonmosaic in regard to sex and the autosomal trait—opaque (dominant) versus transparent (recessive) larval skin—or they were mosaic for sex but not for skin-type, or nonmosaic for sex but mosaic for skin-type, or mosaic for both sex and skin. The explanation for these striking phenomena was obtained both by genetic analysis and cytological observation. Meiosis in insect eggs does not usually

---

FIG. 15. Origin of a mosaic in *Drosophila melanogaster* resulting from double nucleus fertilization. The two egg nuclei carry Sb and not-Sb. They are fertilized by two sperm carrying not-Cy and Cy. (Redrawn from Stern and Sekiguti 1931.)

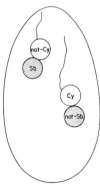

lead to formation of cellular polar bodies but to a series of four nuclei, arranged more or less in a row from the periphery of the egg toward its center. The innermost nucleus is the pronucleus that fuses with the fertilizing sperm. The gene responsible for the inherited mosaicism in the silkworm determines an abnormal arrangement of the polar nuclei so that the pronucleus and its sister nucleus become fertilizable. In Bombyx polyspermy is the rule—it is rare in Drosophila—and two sperm are readily available for the two egg nuclei (Fig. 16). If the mother is heterozygous Aa for the gene involved in the appearance of the larval skin, the two egg nuclei may be either alike, A and A or a and a, or different, A and a. Because femaleness in Bombyx is based on an XY sex-chromosomal constitution, the two nuclei may also either be alike, X and X or Y and Y, or unlike, X and Y. Altogether nine different combinations of the sexual and autosomal genotypes of the two nuclei are possible. If mating has occurred with a homozygous transparent aa male all of whose sperm are X and a, then, according to the type of double-nucleus egg present, any one of the observed kinds of nonmosaic and mosaic individuals is to be expected.

The presence of two genetically different, fertilizable maternal nuclei within one egg may be caused in more than one way. In addition to abnormalities in the fate of polar nuclei, fusion

FIG. 16. (A) Section through an egg of Bombyx mori showing two nuclei fertilized by two sperm. (B) The region of the nuclei, enlarged. The left nucleus of each pair is a sperm nucleus as suggested by the presence of a cytoplasmic "pseudopodium" or of formation of an aster. (A redrawn from Goldschmidt and Katsuki 1928; B from Goldschmidt and Katsuki 1931.)

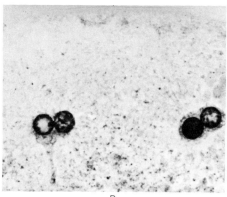

A   B

of oögonia or of oöcytes within the ovary can give rise to single eggs with two nuclei. Such fusions have been studied extensively in the walking stick insect *Carausius morosus* (Cappe de Baillon 1927). Their significance for the production of gynanders has been treated by von Lengerken (1928).

As stated, the mosaicism in *Bombyx mori* depended on a specific genotype responsible for abnormal arrangement of polar nuclei. Another type of inherited mosaicism is known in *Drosophila simulans* (Sturtevant 1929). A recessive gene causes distortion of the first meiotic spindle apparatus, which then leads to dispersion of the chromosomes during anaphase (Wald 1936). Subsequently, failure of the second meiotic division results in the production of an abnormally large number of nuclei, each containing an abnormally small number of chromosomes. Fusion of an X-carrying sperm nucleus with some of the small nuclei without an X-chromosome will result in an XO male genotype, and later inclusion of a "free" maternal X-chromosome in a descendant of the first fusion nucleus will result in a female genotype. This sequence of events accounts well for the fact that most gynanders in this genetic strain exhibit the effects of the paternal X-chromosome in their male parts. It may be added that the gene responsible for the meiotic disturbances also produces a mutant eye color, claret, certainly a peculiar type of multiple genic effect.

Gynanders and Cell Lineage. Sturtevant has used the gynanders produced in the claret stock for a study of cell lineage in development. By suitable X-linked marker genes, such as yellow and singed or forked, which express themselves in any one of the micro- and macrochaetae on the body surface, most of its areas can be assigned unequivocally to the XX or the XO constitution. The majority of the gynanders consisted of approximately half female and half male parts, as would be expected if the gynandric condition were usually established at the formation of the first two embryonic nuclei. An appreciable number of the gynanders were about three-quarters female and one-quarter male or still more preponderantly female. This can be accounted for if the addition of a maternal X-chromosome occurs more than once in successive cleavage stages. The distribution of the two gynandric components in the adult varied greatly from one fly to another: more or less bilateral, anterior-posterior, and many other types similar to what had first been described in Lukas' gynandric bees.

The reason for this variability in gynandric phenotypes lies in the relatively indeterminate embryonic development of insects. A fertilized egg nucleus divides repeatedly without any simultaneous cytoplasmatic division of the egg. Not until several hundred nuclei have been formed in the interior of the egg do they migrate to the surface where they become enclosed in separate cells. The blastoderm formed this way develops only in part into the embryo proper; in part it forms embryonic membranes. Four main variables are involved in the distribution of the nuclear descendants of the first two cleavage nuclei. One is the position of these nuclei within the egg, which depends on the direction of the first cleavage spindle. A second is the variability in the positions and directions of the subsequent cleavage spindles, and a third is the variable route of migration of the nuclei into the blastoderm. The fourth variable depends on the fact that the outside of the adult is derived from relatively few cells of the embryo set aside as imaginal discs. These structures are located below or within the larval body wall and spread out during the pupal stage to form the surface of the fly. It is not surprising then that the distribution of the male and female areas in gynandromorphs reflects the variations in the early history of the nuclei and in the origin of imaginal discs. These variations may even lead to the rare gynanders in which more male than female parts appear on the surface.

Little information is available about the internal organs of *Drosophila* gynanders. The gonads and the genital ducts and their appendages are in a way "self-sexing," and thus provide information on their XX or X constitution. The only other organ for which markers are known are the excretory Malpighian tubules, which are yellow in wild type and colorless in certain recessive genotypes such as the one responsible for white eyes. Nothing is known about markers for sexual differences in the brain. How close is the correlation between the sexual type of the surface of the head and the cerebral complex inside? Problems of cell lineage are not restricted to insects. We shall meet them again when mosaics in man are discussed.

Haploidy and Sex Determination. It was mentioned earlier that the developmental participation of supernumerary sperm or their haploid derivatives which is responsible for most of the haploid-diploid bee gynanders has also been postulated in other organisms. Convincing genetic evidence for the very rare existence of 1n/2n mosaics in *Drosophila melanogaster* was first provided by Bridges (1925, 1930). Later, Crew and Lamy (1938) described such mosaics in the related species *D. pseudoobscura.* These specimens have played an important role in the establishment of the balance theory of sex determination in Drosophila. In 1921 Bridges discovered the so-called triploid intersexes of *D. melanogaster* and found them to possess two X-chromosomes and three sets of autosomes: 2X3A. Since normal females are 2X2A and normal males 1X2A, the intersexual phenotype of 2X3A suggested to Bridges that the X-chromosome carries determiners of femaleness and the autosomes determiners of maleness. Specifically, he assumed that the decisive factor in sex determination is the ratio of X to A. This theory found support in the fact that not only 2X2A diploid flies are females, but so are 3X3A triploids and 4X4A tetraploids, in all of which the ratio X:A is alike and equal to 1. The theory led to the expectation that 1X1A haploid Drosophila individuals also would be female, since in them, again, X:A = 1. This expectation, however, was contrary to the facts in a variety of organisms in which haploidy occurs as a regular phenomenon. Not only in hymenopterous insects like the honeybee and Habrobracon and some other insect families, but also in species of rotifers and mites, where parthenogenesis leads to haploid organisms, the sex of these is invariably male. The maleness of haploid genotypes is shown most dramatically in the coccid genus Icerya (Hughes-Schrader 1927; Hughes-Schrader and Monahan 1966). In these insects two sexual types occur, hermaphrodites and males. The males are haploid and possess only two chromosomes. The hermaphrodites are diploid, having four chromosomes in all somatic tissue and in the ovaries. Initially, the diploid individuals are all female, but early in life some germ cells of the ovary lose one complement of two chromosomes by degeneration and elimination. These haploid germ cells constitute the primary spermatogonia of the hermaphroditic 1n/2n mosaic gonad. The contrast between what was known regarding the maleness of haploids and haploid tissues in so many animal forms and what was predicted for Drosophila was so striking that Schrader and Sturtevant (1923) proposed an alternative to Bridges' formulation of balance that would permit a haploid Drosophila to be a male and thus conform to the general rule.

No purely haploid Drosophila individual has ever been observed, presumably because haploidy results in only very weak specimens which die during development. Fortunately, haploid-diploid Drosophila mosaics are able to survive, as stated above for two species, D. melanogaster and D. pseudoobscura. These 1n/2n specimens proved to be not male-female gynanders but all-female mosaics, contrary to observation in other forms and in agreement with Bridges' theoretical expectation.

Somatic Crossing-over and Mosaicism. The early developmental events that initiate mosaicism naturally lead to large components of each of the different genotypes in the adult animal. It is not rare, however, to find a type of mosaicism in Drosophila in which a small aberrant area occurs on a prevailing background showing the phenotype expected from the initial genotype of the specimen. A female heterozygous for the recessive gene singed (sn/sn^+) will, as expected, have normal, straight bristles, but somewhere there may be a patch of gnarled, singed bristles (Fig. 17A). Or, a female heterozygous for the recessive gene yellow (y/y^+) has, as expected, mostly black bristles, but somewhere there may be one, two, or more yellow bristles forming a patch of aberrant surface tissue (Fig. 17B). And a female whose one X-chromosome carries both yellow and singed and whose other X-chromosome carries the dominant normal alleles of these genes ($y\ sn/y^+sn^+$) may have a patch of yellow and singed bristles on a black straight background. At first, this mosaicism seemed to require nothing more than elimination of one of the female's X-chromosomes from a nucleus at some late developmental stage instead of at the beginning of development, as late, perhaps, in the case of single bristle patches, as the last cell division before the end of ontogeny. The mosaics would be "patch gynanders," overwhelmingly female with a small spot of male tissue. Indeed, some of them may well be of such XX/XO constitution. For others, however, this simple explanation does not suffice. Thus, a $y\ sn/y^+sn^+$ female may have a patch of yellow but not-singed bristles or a patch of singed but not-yellow bristles. Such patches cannot be accounted for by elimination of either one of the X-chromosomes of the fly. For some years after these facts became known there was an inclination to modify the elimination hypothesis to make it account for these partial patches. If only the tip of the normal X-chromosome and not the whole chromosome were eliminated, only the normal allele of yellow would be lost and a yellow patch could appear. If, instead, a middle section of the chromosome were eliminated, the normal allele of singed might be lost and a singed patch

FIG. 17. Somatic segregation in *Drosophila melanogaster*. (A) Side view of a female heterozygous for singed (sn/sn^+). Three singed microchaetae appear on the fourth abdominal segment. (B) Dorsal view of the mesothorax of a female heterozygous for yellow (y/y^+). (C) Dorsal view of the mesothorax of a female heterozygous for yellow and singed in trans arrangement ($y\ sn^+/y^+sn$). Mesothorax B bears a single yellow machrochaeta; C bears a twin spot consisting of 1 yellow macrochaeta and 16 yellow microchaetae, and of two singed macrochaetae and six singed microchaetae.

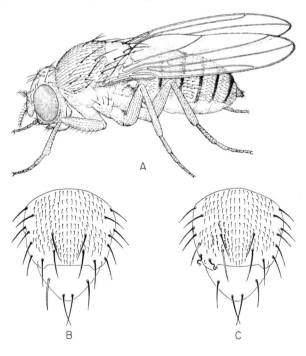

A

B   C

appear. When, however, females were obtained which, while still heterozygous for yellow and singed, carried these recessives not in the same but in opposite chromosomes, that is, in the "trans" arrangement y sn^+/y^+ sn, a new type of mosaicism appeared that called for a basically new explanation. The majority of patches found on the trans-females were twin spots: an area of yellow not-singed bristles next to one of not-yellow singed bristles (Fig. 17C).

It seemed at first as if the twin spots came from two original sister cells into which the two X-chromosomes had been segregated during the preceding division, so that one sister cell received the X-chromosome bearing y sn^+ and the other cell the X-chromosome bearing y^+sn. This would have given gynandric status to the females that bore the twin spots. Detailed analysis soon showed that this was not the case (Stern 1936). The spots were made up of cells which still carried two X-chromosomes, one spot being homozygous y sn^+/y sn^+ and the other homozygous y^+sn/y^+sn. Such constitutions would suggest that the y^+sn chromosome after its usual replication did not succeed in having its daughter chromosomes distributed to separate daughter cells and that the same happened to the replicated y sn^+ chromosome. Yet it seemed that the faulty mechanisms of distribution still worked sufficiently normally to segregate the four chromatids pairwise into the daughter cells—even if they were the wrong pairs. The hypothesis of an imperfect distribution of chromosomes was also applicable to the yellow singed patches found on y sn/y^+sn^+ females, but it failed to explain the rarer yellow not-singed patches and the still rarer not-yellow singed patches on such flies. In the end, it turned out that the distribution mechanism of the chromosomes was not at fault at all, but that the rest of the chromosomes had misbehaved.

The normal distribution mechanism resides in a specialized organelle of each chromosome, the kinetochore or centromere. During a mitotic cycle each kinetochore doubles and the two sister kinetochores move on the spindle to opposite poles. Because each sister kinetochore is part of a replicated sister chromosome, the movements of the kinetochores lead to the inclusion of identical chromosomes in the two daughter nuclei

(Fig. 18A). The misbehavior of the chromosomes in the production of twin spots on the body of a fly consists in their undergoing a process of recombination in somatic cells, a process which in Drosophila is usually restricted to maturing oöcytes. As in meiosis, somatic or mitotic recombination occurs either at replication or during the replicated stage of each chromosome and, at any one level, between only two of the four strands of a pair of homologous chromosomes. Let k_1' and k_1'' represent the sister kinetochores of the original yellow not-singed chromosome and k_2' and k_2'' those of the not-yellow singed chromosome. Let recombination occur between the locus of sn and the kinetochore. Then the following four chromosomes will be ready for mitotic distribution to two daughter nuclei (Fig. 18B):

FIG. 18. Distribution of chromosomes heterozygous for two genes in the trans position. At left, the chromosomes are doubled. The sister kinetochores of the chromosomes are labelled k_1' and k_1'', and k_2' and k_2'', respectively. At right, the chromosomes are distributed to daughter cells. (A) No crossing-over. The daughter cells are alike genetically, and they or their descendents may form not-yellow, not-singed bristles. (B) Crossing-over between the locus of singed and the kinetochores k. The daughter cells differ genetically and may form a twin spot of yellow not-singed and not-yellow singed bristles.

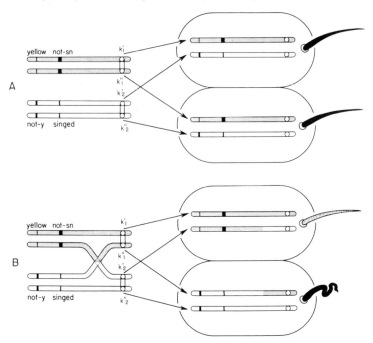

1 yellow not-singed k_1' 3 yellow not-singed k_2'
2 not-yellow singed k_1'' 4 not-yellow singed k_2''

The rules of distribution demand that the two k_1 kinetochores move to opposite poles and that the two k_2 kinetochores do likewise. This permits two alternative types of segregation: (1 plus 4)$\longleftrightarrow$(2 plus 3) and (1 plus 3)$\longleftarrow$ (2 plus 4). The first type of segregation results in a pair of yellow not-singed/not-yellow singed cells that are not different from the rest of the cells. The second type of segregation, however, yields a yellow not-singed/yellow not-singed and a not-yellow singed/not-yellow singed cell, the founding cells of a twin spot.

In Drosophila, mitotic recombination with its subsequent segregation occurs both between the two X-chromosomes and between pairs of homologous autosomes. It provides one of the means for studying the developmental effect of mosaicism in regard to a variety of genes, as will be illustrated in the third essay. Mitotic recombination also occurs in organisms other than Drosophila. It has been extensively investigated in a variety of fungi, including yeast, where colonies heterozygous for the genes determining pigmentation form twin sectors of mitotically segregated homozygous genotypes (Fig. 19; Pontecorvo and Roper 1953; James 1955). How widely mitotic recombination is spread among organisms is not known. It must be assumed that one of the prerequisites of its occurrence is a pairing of homologous chromosomes in somatic cells. In the diptera, somatic pairing is a typical, ever-present phenomenon, and in the fungi it may be present, although the extreme smallness of their chromosomes has not permitted evidence either for or against it. In most other organisms homologous chromosomes are known to pair in meiosis only, but it cannot be excluded a priori that a tendency to somatic pairing may occur in special tissues or, occasionally, in cells of many kinds. Grüneberg (1966) has recently reanalyzed numerous reports on individual rare cases of mosaic coat colors in mice. He comes to the conclusion that somatic crossing-over is the most likely interpretation for a high fraction of these mosaics. L. B. Russell (1964a), on the other hand, finds that most of the mosaicism which appeared in her and W. L. Russell's mouse colony originated as a result of mutation. Students of mammalian cells in tissue cultures are also on the look-out for mitotic recombination, as it would be a valuable tool in the growing field called "somatic cell genetics." Some suggestive cytological findings have been recorded in cultured human cells, and hypotheses involving frequent somatic recombination as a normal process in special tissues of the intact organism have been proposed. These hypotheses will be dealt with below.

Mutational Mosaicism. The various types of mosaics discussed thus far all have one aspect in common. They are compounds of preexisting genetic diversity. There is another type of mosaic which is caused by the occurrence of genetic newness, that is, mutation. A well-analyzed example of such a mutation mosaic is a female mouse which appeared in a carefully bred strain homozygous for the recessive not-agouti fur-color gene *aa* (Bhat 1949). Much of the surface of this individual was not-agouti, like that of all other animals of the strain, but it had a number of irregular patches of agouti-like markings on its back (Fig. 20). It seemed likely that a mutation from *a* to a dominant allele *A* had occurred in a cell of the embryo which thus became *Aa*

Fig. 19. Genetic mosaicism in colonies of diploid yeast cells. The strain used was heterozygous at a locus concerned with adenine metabolism, AD_2/ad_2. Colonies of the heterozygous type are white, those of homozygous ad_2/ad_2 red. The figure shows part of a culture dish with many wholly white and two sectorial white/red colonies that developed from X-irradiated single cells, presumably as a result of mitotic crossing-over. (From Dr. R. K. Mortimer, University of California, Berkeley.)

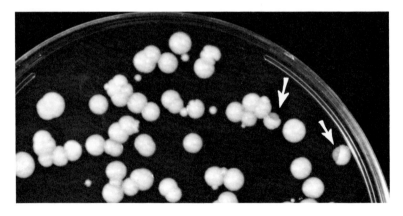

genetically. During subsequent growth and differentiation, descendants of this cell came to be present in the hair follicles of certain regions of the body surface with the result that the fur in these areas became agouti-colored. Direct proof for such an explanation would consist in showing that the pigment-producing cells actually carried a changed gene. This is not yet possible, although analysis of the DNA of such cells might someday be detailed enough to provide the evidence. Very strong indirect proof is available, however, if it can be shown that some of the germ cells of the mosaic actually transmit the mutated gene. This was true in the mosaic-furred mouse specimen. Consecutively mated to two aa males, she produced 61 young of the not-agouti type and 2 young heterozygous for the well-known, dominant agouti allele A^L. Obviously, the mutation had occurred early enough in the life of the embryo so that the descendants of the cell in which it happened gave rise to both germinal and somatic cells. (No theory is exclusive of alternatives! It would be possible that two separate but identical mutations took place during development in this mouse, one in a somatic cell and another in a germ cell. The probability

---

FIG. 20. *Above*, a mouse homozygous for not-agouti (a/a), mosaic by having an agouti-like (A^L/a) patch. The ovary, too, is mosaic as indicated by the outline above the hind leg. *Below*, the fertilized egg and two later embryonic stages in cross sections. It is assumed that a mutation from a to A^L occurred in one nucleus of the blastocyst (*middle*) and that descendants of this cell became incorporated in the embryonic disc (*right*).

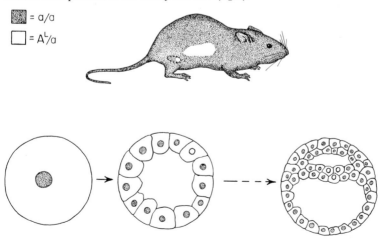

■ = a/a

☐ = A^L/a

of such a double event is presumably much lower than that of the single event postulated above.)

L. B. Russell (1964a) has reviewed over 100 mosaic mice which had appeared in the large-scale studies on spontaneous and radiation induced mutation rates at the Oak Ridge National Laboratory. At least 26 and probably as many as 40 or more were both somatic and germinal mosaics for a new mutation from a dominant to a recessive allele of one or another of five genes affecting coat color. The crosses in which these mutations occurred had been arranged so that the males carried the dominant normal alleles at the loci for which the females were homozygously recessive. The F_1 hybrids may therefore be designated as $AaBbCcDdEe$ and a mutation of a dominant gene, for example a B to a recessive b, would lead to the phenotypically recognizable genotype $AabbCcDdEe$. Obviously only mutations in a paternal gene were ascertainable in these studies. In the 20 mosaics for which the mutational evidence is most decisive, the percent of germinal tissue carrying the mutation varies from as low as 7 to as high as 98, the average being 49.5 percent. If one allows for enough variability in the cell lineage of the descendants of the original mutated cell, one may suspect that in the majority of the mouse mosaics about one-half of all cells of the organism are mutant.

The evidence in Drosophila leads to a similar conclusion (Altenburg and Browning 1961). Many new mutations in the fly appear as "fractionals," affecting only part of the tissue of the individual in which they have first been found; on the whole, they seem to be present in about one-half of all cells. As in the mouse experiments, the experimental procedure consisted of crosses of dominant normal males to recessive females. The ascertainable mutations were those in which a dominant paternal gene had changed into a recessive. The proportion of mosaics among all mutant individuals is very high both after exposure of the male parents to three different kinds of chemical mutagens and in untreated controls in which "spontaneous" mutations occurred. In different series, from 30 to 67 percent of all mutant offspring were mosaic.

These remarkable facts are subject to at least two different interpretations. One can assume, in regard to the spontaneous mosaics, that the fertilizing sperm contained the original, unmutated dominant allele of the male who had produced it. Therefore, the fertilized egg nucleus was heterozygous at each of the loci studied and would usually have developed into an individual of normal appearance. At the first cleavage division, however, or in the two-cell stage, a mutation occasionally occurs with the result that one of the nuclei retains an unmutated gene while the other receives a mutated one. An argument in favor of this interpretation is the fact that, in the mouse, the first cleavage division is unusually sensitive to disturbance of chromosome distribution and that, therefore, it might also be unusually sensitive to mutagenic agents. Against this interpretation it must be said that, in Drosophila, chemical treatment of the male before mating results in increased numbers of mutational mosaics in the offspring. This fact suggests the second interpretation of the mutational origin of mosaics. It assumes that an abnormal sperm is involved in fertilization instead of a normal sperm whose normal gene mutates later.

The ability of an abnormal sperm to transmit to an embryo both a nonmutated normal and a mutated new gene at a given locus is best explained by some preexisting doubleness of the gene. Such doubleness is inherent in the double-stranded structure of the DNA molecule. If a mutation in a sperm initially consists in a change at the position of only one member of a nucleotide pair, then the replication of the two strands after fertilization may result in two different alleles of the same gene. The changed strand may be paired complementary with a correspondingly changed newly synthesized strand resulting in a stable mutant gene. The unchanged strand will be paired with the usual newly synthesized strand resulting in a stable normal gene. In detail these molecular events may be of many different kinds. Some of them will also account for mosaics in which only one-quarter of the resulting mosaic individual will be mutant, and others may result in permanently unstable genes causing mosaicism in later cell generations, including those representing whole individuals of later generations.

Mutational Mosaicism in Man. There are many mosaics, in mammals as well as in insects, which do not transmit any genetic change to their offspring. Undoubtedly, many of these mosaics nevertheless are mutational in origin. Direct evidence for mutation depends on transmission of the mutated gene to another generation. Absence of transmission is not proof against mutation, but may often be the result of the circumstances that no cells of the germinal tissue happened to receive the mutant gene. The opposite situation exists when the soma of an animal is homogeneously nonmutant as far as can be discerned, but its gonads contain mutated germ cells. If a mutation occurs as late as in an oöcyte which will mature into a single egg, or if it occurs in a sperm cell, one can hardly speak of a mosaic gonad. When, on the other hand, numerous germ cells of a gonad are derived from one ancestral mutated cell, then one encounters true internal mosaicism. Such an event may have occurred in a human family in which a normal couple had six normal children and four children in whom the iris of each eye was absent (aniridia; Reed and Falls 1955). For three generations this trait was transmitted to 19 descendants in the dominant fashion characteristic of aniridia in other pedigrees. Why then was neither of the original parents affected? The answer likely lies in an early embryonic mutation in a cell of the prospective mother or father. The descendants of this cell did not include those somatic cells which would be responsible for absence of the irises, although they may have been present in other parts of the body than the eye where their effect is not recognizable. Some of them, however, developed into germ cells which subsequently betrayed their mutant character in the four affected children.

The possibility of gonadal mosaicism may weigh heavily on the human genetics counselor. Assume that he is called to advise a normal couple who have an abnormal child about the chances of recurrence of the abnormality in later pregnancies. Assume that he comes to the conclusion that a dominant newly mutated gene in one of the gametes of the parents led to the child's abnormal development. He may be inclined to reason that the chance of such a mutation 's 1 in 10,000 or less, and that therefore the chance of another one is equally low. This is probably true in the majority of families in which the first mutation occurred in a single terminal germ cell. Yet there must be persons with mosaic gonads in whom the chance of recurrence would be much higher, even approaching 50 percent. We do not know what the relative frequency of single terminal mutations to earlier mutations is in man. While it has been customary to think of mosaic mutational human gonads as rather rare, this opinion may well have to be revised in the light of the high frequency of such mosaics in flies and mice. Be this as it may, had a counselor been advising the parents of the aniridic children at the time when only one of them had been born he might have been pardoned if he judged as minimal the possibility of a second affected child. Had he been consulted after the second affected was born, he would surely have had to consider that gonadal mosaicism existed in one of the parents, and his prognosis for another recurrence would have had to be more pessimistic.

Chromosomal Mosaicism in Man. By the nineteen-thirties, the cytological and genetic interpretations of mosaicism had, on the whole, been well established. They had been mostly derived from the study of invertebrates, particularly insects. Their application to vertebrates had not gone far, for a variety of reasons. Fewer large-scale genetic experiments had been conducted with vertebrates than with invertebrates, and fewer genetic markers demonstrable in mosaics were known. As to sex mosaics, the dependence of the higher vertebrates on hormones for sexual differentiation seemed to make it unlikely that gynanders, if they occurred, would be recognizable. Nevertheless, Morgan and Bridges (1919), who surveyed a large number of reports on gynanders among both invertebrates and vertebrates, including man, expressed the opinion that many of these individuals may have been genetic mosaics of the type so convincingly demonstrated in Drosophila. Later, similar suggestions were made for certain specimens of birds (Keck 1934; Danforth and Price 1935; Danforth 1937a, b; Crew and Munro 1938) and mice (Hollander, Gowen, and Stadler 1956). On the whole, however, and particularly in human "hermaphroditism," understanding concerning genetic aspects remained rudimentary.

This changed dramatically after the midcentury, when the students of mammalian tissue cultures succeeded in greatly improving the methods of analyzing the chromosomes of cultured cells. Better spread of the chromosomes in mitotic preparations and larger numbers of suitable stages permitted accurate chromosome counts and detailed characterization of many individual chromosomes. For man, these technical advances resulted in 1956 in the recognition of 46 (and not 48) as the number of chromosomes in normal diploid human cells (Tjio and Levan 1956). At first sight this might have seemed to be a minor descriptive detail, but in proved to be the necessary basis for the next and most significant discoveries. These were the findings that various well-known types of abnormal human individuals carry abnormal chromosome numbers or abnormal chromosome configurations, characteristic for each type.

The study of mosaics in insects was carried on nearly exclusively by means of genetic markers. Direct, cytological proof of chromosomal diversity within an individual usually cannot be obtained by observations on somatic tissues, because no cell divisions occur in most parts of the adult insect. Only in those rare instances when cells of the gonads are involved can chromosomal mosaicism be demonstrated independent of genetic data. This has been possible in several somatically and gonadally gynandric grasshoppers (Amblycorypha). As predicted, ovarial tissue contained two X-chromosomes and its neighboring testicular tissue only one X (Pearson 1927). In Drosophila the only cytological study of a mosaic was made in a species hybrid (MacKnight 1937). The animal was a female composed of normal diploid and seemingly haploid tissues. Indeed, of two analyzable cells in one of its ovaries, one cell was diploid and the other haploid.

In contrast to the situation in insects, cytological study of human mosaics has yielded many more discoveries than did genetic analysis. An insect usually has to be killed to prepare its few possibly informative cells for chromosome analysis. A human individual can harmlessly provide repeated biopsies of bone marrow cells, blood cells, skin, and other tissues in which the chromosomes can be studied in short- or long-term tissue cultures.

One of the first-described chromosomally aberrant human beings was a mosaic (Ford et al. 1959). This man had the characteristics of the Klinefelter syndrome, which had just become known as typically based on an XXY constitution. His bone marrow cells, however, were of two types, XXY and XX.

Chromosomally, he thus was a gynander although his phenotype was essentially male. It remains, of course, unknown, from a study of bone marrow cells alone, how much of the mosaicism pervades the whole body of an individual. Bone marrow studies along with studies on circulating blood yield somewhat more information and, in addition, fibroblasts from biopsies of skin from different regions of the body allow still further insight into the extent of the mosaicism. Even biopsies of internal tissues may at times be feasible when surgery is called for in the treatment of abnormalities. It is unavoidable, however, that some mosaics remain undiscovered, because even extensive sampling of cells from different body parts is liable to yield only one of the chromosomal constitutions and miss another. To some extent this depends upon the stage of development at which the mosaicism originated. The later the stage, the more likely it is that major parts of the body will be of only one chromosomal type, that present at the beginning of development. The earlier the second karyotype is initiated, the more evenly distributed will the two cell types be. Much variation in detail can be expected, depending on cell lineage relations, cell migration, and chance phenomena.

There is only one study in which the distribution of two chromosomal types in a whole mosaic fetus has been recorded (Klinger and Schwarzacher 1962). This was a conceptus with crown-rump length of 60 mm, corresponding to the twelfth to thirteenth week of pregnancy. It was removed by Caesarean section, in a Swiss hospital, for maternal psychiatric indications that would not be expected to affect fetal development. A study then in progress at the University of Basel required tissue samples for culture from several regions of the body of a fetus and the aborted specimen was used for this purpose. In addition, the fetus, its placenta, and its membranes were preserved after histological fixation. When it was discovered that the fetus was a mosaic, it was serially sectioned for detailed microscopic analysis and its amnion and part of its chorion cut in portions and also mounted on slides. The mosaicism of this fetus was demonstrated decisively by analysis of the chromosomes of dividing cells. Nearly 700 such cells from tissue cultures of various organs were inspected. Two-thirds of these cells had

FIG. 21. Distribution of cells with sex chromatin in an XY/XXY mosaic human embryo. (A) A portion of the amnion epithelium. Left half, a sex chromatin positive area; right half, a negative area. (B) Transverse section of the esophagus. (C) Tracing of the section with information on the incidence of sex chromatin in the epithelial layer. (From Klinger and Schwarzacher 1962.)

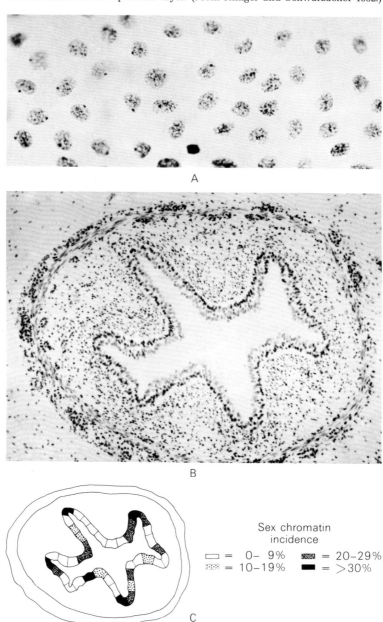

47 chromosomes and one-third 46, and karyotype analysis of some 70 cells revealed that the prevailing mosaicism was due to presence of XXY and XY cells.

The majority of cells both of the fetus and of the tissue cultures could not be classified on the basis of chromosomal counts because they were in interphase, but a useful criterion of sex-chromosomal constitution is provided by the sex chromatin. This is a small, densely-staining body discovered by Barr and Bertram (1949) in the interphase nuclei of many mammalian cells which have two X-chromosomes, but which is present in only a very small percent of cells with a single X-chromosome. The incidence of sex chromatin in various tissues of the XY/XXY mosaic averaged about 10 percent. Thus it was higher than in whole XY and lower than in whole XXY persons. The average alone gives only limited information. The variations in incidence of the "Barr bodies" were not distributed according to chance. Instead, cells with Barr bodies occurred in clusters and the same was true for cells without them (Fig. 21). Counts of tens of thousands of cells in situ and in cultures showed that the mosaicism occurred in all tissues: amnion, chorion, epidermis, esophagus, liver, lung, brain, and elsewhere. The mosaic pattern occurred on a "micro" scale throughout the entire body. No single organ or region contained nuclei of one type only.

Sex-chromosomal Mosaics. Many types of sex-chromosomal mosaicism have been recorded (Fig. 22). Some mosaics like the

---

FIG. 22. A variety of sex-chromosomal mosaics found in man. Phenotypically the mosaics may be female, male, or gynandric.

Human Sex-Chromosome Mosaics

female	male	gynandric
XO / XX	XY / XXY	XO / XY
XO / XXX	XY / XXXY	XO / XYY
XX / XXX	XXXY / XXXXY	XO / XXY
XXX / XXXX	XY / XXY / ?XXYY	XX / XY
XO / XX / XXX	XXXY / XXXXY / XXXXXX	XX / XXY
XX / XXX / XXXX		XX / XXYY
		XO / XX / XY
		XO / XY / XXY
		XX / XXY / XXYYY

XY/XXY fetus are male throughout; some, like the XO/XX individuals, are female throughout; and others such as the earlier described XX/XXY individual are at least formally sexual mosaics, even if single-sexed in general appearance. Triple mosaics are also known, such as XO/XY/XXY. Such types of chromosomal diversity within an individual are the result of a variety of processes. For some mosaics the assumption of a single event of chromosome elimination or of abnormal distribution of chromosomes in mitosis is sufficient as, for instance, shown in the diagram depicting the possible origin of an XO/XY gynander (Fig. 23). For other mosaics two or even more separate abnormal events have to be assumed, as for example in an XO/XY/XXY mosaic where a presumably XY zygote lost the Y-chromosome from one stem cell and gained an X-chromosome in another. Some multiple events may involve both prezygotic and zygotic stages. Thus, an XY/XXY mosaic may have started as an XXY zygote which came from a normal X-carrying egg and a non-disjunctional XY sperm. After fertilization one of the two X-chromosomes may have been lost from a cell, resulting in the XY lineage of the mosaic. Here, as in most other mosaics, more than one interpretation may fit the observations. One alternative would be to assume that the mosaic started out as an XXY zygote which subsequently lost one of its X-chromosomes from a cell and later its Y-chromosome in a descendant of the XY cell.

It might be wondered how probable it is that two or more rare cytological events coincide. On the one side there is evi-

---

FIG. 23. One possible mode of origin of a human gynander. (A) Fertilization of an egg by a Y sperm. (B) First cleavage division entailing loss of a Y-chromosome. (C) The resulting two-cell stage consisting of an XY and an XO cell. (D) Cross section through blastocyst with XY and XO cells. (From Stern, 1960c.)

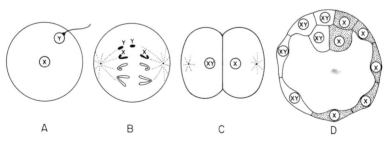

A   B   C   D

dence for the existence of genetic and nongenetic agents which express themselves by general increases in the appearance of chromosomal abnormalities. On the other side, even if the events are independent of each other, the probability of their occurring together, while certainly very low, is not so low as to exclude the possibility of detection. It must be remembered that mosaics are selected for investigation from hundreds of millions of human beings.

The great variability of human hermaphrodites which so puzzled investigators until recent times can now be understood as the consequence of gynandric mosaicism. The origin of a sex-chromosome constitution different from that of the fertilized egg may occur early or late in development. Accordingly, the mixtures of genotypes may be present in every organ or result in a coarse mosaic. The variability of later cell lineages is superimposed on the variability of initiation of the mosaic condition. Most important, and underlying all external and internal expression of sex mosaicism, is the apparent ability of male and female tissues and cells to react differentially to the same assortment of sex hormones that is diffused and circulated within the body.

Human gynanders in whom sex-chromosomal mixtures may result in mosaicism of the sexual organs, ducts, and gonads may have severe psychological problems (Money 1963). Human self-identification as a male or female seems to depend on a complex of circumstances. It is probable that at birth no psychosexual differentiation has occurred. When it appears later, it is strongly influenced by the social environment that leads to the rearing of the child as either a male or a female. In the overwhelming majority of individuals, this assignment of sexual type by the outside world is based on normal male or female genital morphology and no conflict between assignment and body image exists. In the small group of individuals in which mosaicism or other circumstances result in an ambiguous morphology, self-identification may be either inclined toward the assigned sex of rearing or, probably more rarely, to a repudiation of assigned sex. How much, in addition, sexual self-identification depends on or is influenced directly or indirectly by genetic differences in psychological reactions is unknown. Sometimes surgery is applied to gynanders with the primary goal of making the physical features of the body conform more closely to the existing sexual self-identification of the individual. Such surgical procedures may be used to "reverse the sex" of a gynander in terms of social assignment.

It is tempting to quote the extreme case of "testicular feminization" (Morris and Mahesh 1963). Persons with this genetically conditioned complex of traits have the XY-chromosomal constitution of a male throughout their tissues, and their gonads are testes. Otherwise their body is that of a female except for the absence of a uterus and of oviducts. Notwithstanding their XY constitution, these individuals regard themselves as women and frequently marry. This seems a highly suggestive example for arguing the absence of correlation between "chromosomal sex" and psychological sex identification, but there is a possible loophole in the argument. The gene which is responsible for testicular feminization induces the body cells, in spite of the presence of a normal Y-chromosome, to react during growth and development in a female way. The possibility cannot be excluded—even though it may be unlikely—that the XY brain cells of these persons, just as the rest of their tissues, react in a fashion resulting in female self-identification.

There is an interesting account of the mating behavior of a rat whose reproductive anatomy suggests that it was a gynander (Beach 1963). Its external and internal genital structures were composed of female and male elements including an ovary on one side and an ovotestis on the other. When confronted with male rats the mosaic animal evoked copulatory behavior of the males and responded to them in female fashion. When placed with females the mosaic often reacted in masculine fashion. Clearly the bisexual mating behavior of a mosaic conformed to its bisexual anatomical makeup.

Mosaics Involving Autosomes. Human chromosomal mosaicism is not restricted to the sex chromosomes. Mosaicism for a single human autosome exists in some persons with features of Down's syndrome (mongolism; Lejeune 1964, review). Most affected individuals are nonmosaic; they possess 47 instead of 46 chromosomes because of the presence of three instead of two homologous small chromosomes designated as number 21 of the chromosome set (21-trisomy). Some persons are triplo-

21/diplo-21 mosaics, and one individual was found whose blood contained a mixture of three cell types having 46, 47, and 48 chromosomes. Specifically, these cells were diplo-21, triplo-21, and quadruple-21 with frequencies of 42, 53, and 5 percent, respectively. Again, a variety of known irregularities in chromosome behavior can account for the 21-mosaicism. Specific interest in these persons is focused on the association or lack of it between the physical features and brain function as reflected in mental ability. The known 21-trisomy mosaics include one with rather normal appearance but very low intelligence and one with average intelligence though typical physical stigmata. Postmortem examinations of the brains of presumably nonmosaic patients with Down's syndrome have shown a number of relatively unspecific abnormalities. Study of the brain of mosaics may perhaps throw new light on the neurological aspects of the disease.

One of the most unexpected kinds of human mosaicism is found in individuals whose tissues are mixtures of cells with 46 and 69 chromosomes. They are diploid/triploid mosaics or "mixoploids" (Böök 1964). The triploid condition may be due to the initial presence of either an unreduced diploid egg nucleus or a diploid sperm nucleus, or perhaps it is initiated by triple fusion of a haploid egg nucleus, a haploid sperm nucleus, and another haploid egg nucleus which would normally have been destined to be included in a polar body. On the basis of cytological findings on polyspermy and triploidy in various mammalian embryos (Austin 1960), even the fusion of a haploid egg nucleus and two sperm nuclei resulting in triploidy has been suggested. Still other possible variants for the production of a triploid zygote can be envisaged.

If triploidy is the primary event, the origin of 3n/2n mosaicism would be a secondary phenomenon consisting of the elimination of one chromosome set from a 3n nucleus. One would have to think of this elimination perhaps not so much as a single step mechanism, though this is not excluded (Gläss, 1957, 1961), but as the result of a succession of abnormal mitoses which gradually reduce the 3n to the 2n condition. The intermediate aneuploid cells would be expected to be genically unbalanced and of low viability, but the diploid state, once attained, would result in normal survival if not in a selective advantage over the triploid cells. This sequence of events was suggested for a triploid/diploid/aneuploid mosaic rabbit, studied as early as 1950 by Melander. Similar stepwise mitotic reduction from diploid to haploid nuclei is a not infrequent process in the mold *Aspergillus nidulans* (Käfer 1961). In this connection it is of interest that in the first described 3n/2n mosaic human individual, the proportion of triploid cells decreased during the period of observation from 84 percent in skin cells at the age of 14 months to 10 percent at $4\frac{1}{2}$ years, and from 49 percent in connective tissue cells to 7 percent.

Triploid/diploid mosaicism may not, of course, have been derived from a triploid zygote but, on the contrary, been produced from a fertilized diploid egg. Böök inclines to this view. He suggests that normal fusion between egg and sperm nucleus was followed by a mitotic division into two nuclei, one of which combined with a retained polar-body nucleus, thus resulting in one diploid and one triploid cell. Still other possibilities for the initation of mixoploidy have been discussed by Chu, Thuline, and Norby (1964) in a paper on a 3n/2n mosaic male tortoiseshell cat.

Nonidentical Monozygotic Twins. Chromosomally mosaic individuals are exceptions to the rule that the process of mitosis assures identical chromosomal constitution in all somatic cells. In a special form this rule should also apply to monozygotic human twins who, in a way, result from "accidental" separation into two parts of the embryonic cells derived from a single zygote. It is in conformity with this expectation that such "identical" twins are same-sexed and that, if one is affected with a chromosomal abnormality as, for instance, that responsible for Down's syndrome, the other is also affected. A few exceptions to these nearly universally met expectations are now known to exist. Two special pairs of twins, in whom one partner is female and the other male, have come to attention. In the first pair, studies of the blood groups resulted in a probability higher than 99 percent that the twins were of monozygotic origin, a conclusion which was practically proven by the fact that reciprocal skin grafts between the twins survived fully as if they were autografts (Turpin et al. 1961; Lejeune and Turpin 1961). Why then were these twins of different sex?

The answer came from a cytological analysis. While the boy had a normal XY complement, his twin sister was XO. Thus this pair of twins corresponds to the above described XO/XY gynanders, except that early separation of the XY and XO cells of a mosaic embryo resulted in two nonmosaic persons of opposite sex.

Discordance of identical twins in regard to Down's syndrome has also been established by a group of French authors (Lejeune et al. 1962). The identity of each of numerous tested blood antigens, serum components, and other properties of a pair of male twins strongly suggests monozygosity, but chromosomally one was normal, the other 21-trisomic. This "heterokaryotic monozygosity" may have developed from a trisomic zygote which during cleavage lost the extra chromosome 21 from a cell that ultimately split off to form the normal twin. Alternatively, and perhaps more likely, the zygote started as a typical 21-disomic which during an abnormal cleavage mitosis gained an extra chromosome. Here too the rare incidence of abnormal chromosome distribution was accompanied or followed by a twinning process.

One more pair of monozygotic twins deserves to be noted (Mikkelsen, Frøland, and Ellebjerg 1963). Blood grouping of a pair of twin sisters gave a 98-percent probability that they were identical, but one of the twins, Elsa, suffered from severe mental deficiency and had signs of Turner's syndrome while the other twin, Karin, was phenotypically and mentally normal. Both twins, it was found in blood and skin tests, were XO/XX mosaics. No sex chromatin, however, was found in buccal smears of Elsa, which suggests that only XO cells were present in the epithelium of the mouth cavity; in contrast, sex-chromatin positive cells constituted about 25 percent of the cells in buccal smears of Karin. These twins are remarkable in their great dissimilarity despite possession of the same kind of mosaic constitution. Nevertheless, the distributions of the two nuclear components may well have been very different in the two sisters, as suggested by the difference in sex-chromatin findings and by the fact that chromosome counts on blood and skin cultures in Karin gave similar frequencies of 45 and 46 chromosome nuclei, but those on Elsa yielded highly significantly different proportions. Differences such as these may well result in the normal appearance of one twin and Turner-like abnormality in the other. They do not, however, account for the mental deficiency in one twin, which still is not understood. As to the origin of the mosaicism, its occurrence may have preceded the twinning event in an initially XX blastomere, or loss of an X-chromosome may have taken place in a cell of each embryo after separation of still normal cleavage cells. The same kind of alternatives would apply if the zygote started as XO and a subsequent abnormal mitosis resulted in XX cells.

Late-originating Mosaicism. The chromosomal mosaicism described in the preceding cases was of a rather coarse nature, the frequencies of the varying nuclear types being of the same order of magnitude. One may assume, therefore, that the nuclear differences became established early in the life of the mosaic individuals. The abnormalities in chromosome distribution that lead to mosaicism are, of course, not restricted to embryonic stages. In most tissues, mitoses occur all through life, and the steady wearing-out of the 25 trillion (2.5×10^{12}) circulating red blood cells of a person (see Knudson 1965) requires an immense number of cell divisions, any one of which has some chance of being abnormal. The result is that all human beings carry a certain number of chromosomally aberrant cells and that most counts of dividing cells reveal a small fraction with one or more chromosomes above or below that of the great majority. In general, no special significance seems to be attached to this late-originating mosaicism, but there are at least two phenomena which deserve mention. One of these is the fact that the frequency of cultured blood cells with 45 chromosomes increases with the age of the person (Court Brown et al. 1966). In normal males this is the result of loss of the Y-chromosome from some of their cells, and in normal females of loss of an X-chromosome, thus resulting in XO cells in both sexes. In males this increase of XO cells is probably not apparent until the age of 65 years and never results in more than 2 percent of such cells. In females it is apparent a decade earlier and reaches a frequency of about 7 percent. As stated, these findings were made on cells which had been kept outside the body in blood cultures. It remained to be determined whether the increase in XO cells actually occurs in the body or in the cultured cells only. An analysis of their data on the relation between frequencies of XO cells in short-term and longer-term cultures led Court Brown and his associates to the conclusion that the increase in XO cells with age actually occurs within the body, but that their frequencies are enhanced by prolonged culture. One wonders whether the observations give a clue to chromosomal mechanisms as causes or as secondary effects of cellular aging processes.

The second important instance of late-originating mosaicism concerns the so-called Philadelphia chromosome (Ph[1]), a small chromosome which seems to be a fragment constituting about one-half of the long arm of chromosome 21. Presence of the Ph[1] chromosome is correlated with chronic myeloid leukemia. Before onset of the disease the leukocytes and bone-marrow cells of the individual have two normal chromosomes 21 and no Ph[1] chromosome. With the development of the affliction, increasing numbers of cells occur in which one of the typical chromosomes 21 is replaced by the fragment (Fig. 24; Nowell and Hungerford 1961; Baikie et al. 1960). There seems to be little doubt that the cells containing Ph[1] behave as if they are the leukemic cells but, as so often, there is still no final agreement as to what is cause and what effect.

The mosaicism of normal cells and those containing Ph[1] which is characteristic for chronic myeloid leukemia may include more than the two cell types described above. Court Brown (1965) has reported on a patient who had four cell lines in his blood, an XY Ph[1]-positive line and an XY normal line, an XXY Ph[1]-positive line and an XXY normal line. One may assume that the Ph[1] chromosome originated in a single XY cell followed by nondisjunction of the X in a descendent cell. Other schemes that could account for the initiation of the four lines can also be designed.

Human Mosaics from Double Fertilizations. The human mosaics discussed in the preceding pages were primarily discovered by cytological methods. At present these methods mostly provide information on mosaicism in numbers of chromosomes, particularly when the sex chromosomes or chromosome 21 are involved. In certain situations genetic methods, even though less direct than cytological ones, are more powerful because they

FIG. 24. Metaphase plates of two dividing cells from peripheral blood culture of a male patient afflicted with chronic myeloid leukemia. (A) A normal cell with a Y-chromosome and 4 typical small chromosomes belonging to numbers 21 and 22. (B) A cell in which one of the small chromosomes is represented by the shortened chromosome Ph[1] (arrow). The Y-chromosome and the 4 small chromosomes of each cell are illustrated next to each photograph. (A, B from Nowell and Hungerford 1961; the drawings after photographs of these authors).

may yield data on a great variety of loci distributed over many chromosomes. Some human mosaics are now known for whom combined genetic and cytological analyses have proven that double fertilizations were involved in their conception. We shall describe in some detail the first two cases of this that were found.

One of them was a girl, two years old, who came to medical attention on account of an enlarged clitoris (Gartler, Waxman, and Giblett 1962). A study of her chromosomes showed her to be mosaic for XX and XY constitutions. On the average, about equal numbers of cells with these two karyotypes were found in cultures from blood, skin, and internal organs, but there was some topographical segregation of the two cell types. Surgery revealed an ovary on one side and an ovotestis on the other side of her body. One of the eyes of the child was hazel, the other brown. This suggested that the mosaicism was not restricted to the sex chromosomes, and a search for autosomal markers of possible mosaicism was initiated. Such markers are available in the form of the numerous blood-group genes, most of which express themselves individually in the red blood cells that bear them (or, more accurately, that bear them before the normal degeneration of their respective nuclei). The immunological tests gave unequivocal evidence for the presence of two populations of erythrocytes in the child. Leaving aside the genotypes for two blood groups that did not add to the relevant information, one population of red cells was MS^u/MS; R^1/R^2; the other MS^u/Ns; R^1/r (Fig. 25). (MS^u, MS, and Ns are alleles at the L^{MS} locus, R, R^2, and r alleles at the R locus. Each allele produces a specific array of immunological effects). The mother of the mosaic had blood which indicated her genotype to be

---

FIG. 25. Genotypes of the blood groups of two parents, their mosaic child, and the parental egg nucleus (or nuclei) and sperm cells involved. One of the sperm carried an X-, the other a Y-chromosome. (Based on the data of Gartler, Waxman, and Giblett 1962.)

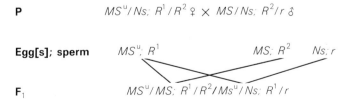

P MS^u/Ns; R^1/R^2 ♀ × MS/Ns; R^2/r ♂

Egg[s]; sperm MS^u; R^1 MS; R^2 Ns; r

F_1 MS^u/MS; $R^1/R^2/Ms^u/Ns$; R^1/r

$MS^u/Ns;$ R^1/R^2. The father had the genotype $MS/Ns;$ R^2/r. Paternally the child's blood-cell constitutions must have been derived from two sperm, one carrying $MS;$ R^2, the other $Ns;$ r. In order to result in the gynandrism of the child, one of these sperm must have been X- and the other Y- bearing. The mother's contribution consisted of only one genotype, $MS^u;$ R^1.

The initial events in the origin of the mosaic child cannot be deduced unequivocally. The mother was heterozygous for all four tested blood- and serum-group loci, but the child received from her only one of the $2^4 = 16$ possible allele combinations. Should one then assume that a single haploid egg divided into two blastomeres before each blastomere nucleus fused with one of the two sperm respectively (Fig. 26A)? Or could two separate eggs have been fertilized by the two sperm and the two products later have fused to form a single embryo (Fig. 26B)? In this case, only one chance in 16 would have been realized when both eggs received identical maternal genotypes. There are still other processes that could have given rise to the mosaic individual, such as double fertilization of an egg and one of its presumably abnormally large polar bodies. The important situation remains of a human individual derived from a fertilization process involving two separate sperm.

The second person of such unusual mosaic nature provided evidence not only of the involvement of two sperm cells in his

FIG. 26. Two possible modes of origin of the human mosaic whose genotypes were shown in Fig. 25. (A) Two sperm fertilize two identical nuclei of the first two blastomeres. (B) Two sperm fertilize two eggs which later fuse to form a single embryo.

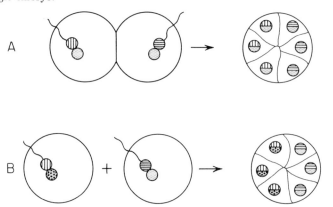

origin but also of two different maternal nuclei (Zuelzer, Beattie, and Reisman 1964). The subject was an 18-year-old male of Negro-American Indian-Caucasian ancestry who was first studied as a prospective blood donor. It was found that the majority of his blood cells belong to group A, but that a minor population of group B cells, amounting to about one-tenth of all cells, also was present. Other traits related to genetic properties of the blood, involving the Kidd factors, secretor status (Se-se), a related anomaly of the Lewis system, as well as hemoglobin types also occurred in two varieties, one represented by the major and the other by the minor population. In addition there was a mosaicism in skin pigmentation with patches of relatively dark brown coloration on a background of very light, café-au-lait color. Chromosome analysis showed a gynandric condition of predominantly XY cells compounded with XX cells. Had it not been for the "accidental" discovery of the blood mosaicism of this individual, he would presumably not have come to attention as a sex-chromosomal mosaic. Apart from a slight enlargement of the mammary glands he was externally a normal male, and biopsies of his gonads showed them to be testes with active spermatogenesis.

The mosaicism of this individual was based on two different genetic contributions from each of his parents (Fig. 27). The father was AB and contributed the A allele in one sperm to his son's major population of cells and the B allele in another sperm to the minor population. The father was also Se/se, and contributed the Se allele together with A and the se allele with B. One of the two spermatozoa involved was X-containing, the other Y-containing. The mother was BO and se/se and contributed O to the major A-carrying population, while her corre-

---

FIG. 27. Genotypes of two parents, their mosaic child, and the parental egg nuclei and sperm cells involved. One of the sperm carried an X-, the other a Y-chromosome. (Based on the data of Zuelzer, Beattie, and Reisman 1964.)

P $B/O; se/se; Hb^S/Hb^A$ ♀ × $A/B; Se/se; Hb^A/Hb^A$ ♂

Egg[s]; sperm $O; se; Hb^S$ O or $B; se; Hb^A$ $A; Se; Hb^A$ $B; se; Hb^A$

F$_1$ $A/O; Se/se; Hb^S/Hb^A / B/-; se/se; Hb^A/Hb^A$

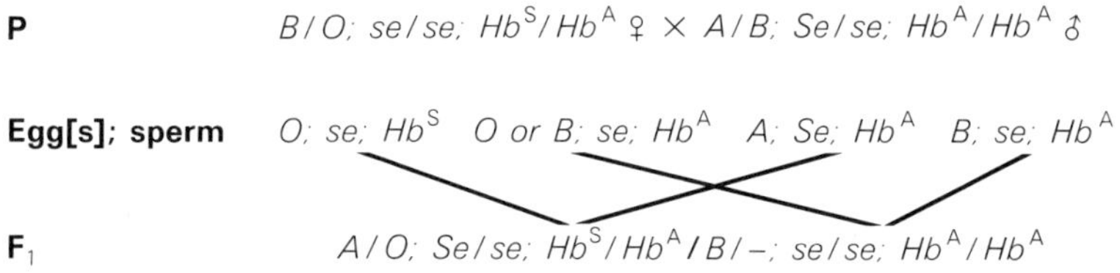

sponding contribution to the minor, B-carrying population remained unknown. She was also heterozygous for the sickle-cell hemoglobin gene Hb^S/Hb^A and contributed, separately, both the Hb^S and the Hb^A allele. Specifically, the A Se containing cells of the son were Hb^S/Hb^A, and his B se/se containing cells Hb^A/Hb^A.

It is clear that two egg nuclei were involved in the origin of this mosaic person, each one fertilized by a separate sperm. Zuelzer and his colleagues are inclined to interpret their findings, particularly in regard to the pronounced quantitative disparity of the two fertilization products, as evidence of double fertilization of a maternal gamete possessing two dissimilar nuclei representing an egg and its smaller polar body. This is indeed a possible origin, but fusion of the cleavage stages of two separate fertilized eggs seems not to be excluded. Differential rates of development and unequal cell-lineage distributions of the two initially separate zygotic products might equally account for the difference in quantity of the two kinds of cells present in the fully developed mosaic.

Double-fertilization Mosaics in Mice. The first human double-fertilization mosaic was described in 1962 and the second in 1964. Within that short period a corresponding mosaic mouse was discovered (Woodiel and L. B. Russell 1963; L. B. Russell 1964*a*). It was a female with yellow and black coat areas arranged roughly in transverse bands. The animal was fertile and was bred intensively. Instead of the maximum of two alleles which a normal diploid animal can transmit, she transmitted three alleles of the A locus for pigmentation, A^y, a^x, and a. Her mother was heterozygous A^y/a^x and her father homozygous a/a. Clearly, the mother contributed her two alleles and the father the third allele to their mosaic daughter. She had the normal number of 40 chromosomes in both yellow and black areas, thus excluding the possibility that chromosome 5, the bearer of the A locus, was present in trisomic state. It appears then that this mosaic mouse had two kinds of genotypes, A^y/a (yellow patches) and a^x/a (black patches), a supposition that was strengthened by tests for the segregation of linked markers in her offspring. Russell suggests that this mosaic resulted from fertilization by two identical sperm of an egg and its first or second polar body, but notes that if the animal had been formed by fusion of two originally independent zygotes or embryos, the results would be indistinguishable from those obtained.

It has been pointed out repeatedly in the preceding pages that it is not possible at the present time to decide between the various proposals regarding the origin of the double-fertilization mosaic individuals, mouse and human. It appears, though, that there is little or no precedent in vertebrates to support the hypotheses suggesting involvement of a single egg, either after parthenogenetic division or including one of its polar bodies. There is, however, experimental evidence for creation of a single individual from fusion of two early embryos. Such animals were first produced in newts, either within one species or heteroplastically as combinations of two species (Mangold and Seidel 1927). More recently, mosaic mice have been produced by fusing two developing eggs during early or mid-cleavage (Tarkowsky 1961, 1963, 1964a, b; Mintz 1965). The fusion was accomplished outside the mothers' bodies and the compound embryos transferred afterward to the uterus of pseudopregnant foster mothers. The resulting fully developed mice were delivered by Caesarean section or born naturally. By using fertilized eggs from strains different in their pigmentation or from crosses yielding pigmentation differences, the mosaic nature of the animals could be analyzed in detail. In addition to the pigmentation mosaicism, sex mosaicism was observed. Some of the mice were normal females or males, and others were gynanders, as expected from fusion of male and female embryos. Undoubtedly these mosaic mice will be very important for the study of various problems in developmental genetics and of the role of the gonads and other organs in embryonic and postnatal sex differentiation. Apart from these fundamental implications of artificial mosaicism in mice, it is worth noting with Mintz that the fusion products of two different types of first-generation hybrid embryos grew up into individuals, each of whom had four parents!

Blood Mosaics in Cattle and Man. In 1945, Ray Owen showed that the majority of pairs of genetically nonidentical, dizygotic cattle twins have identical erythrocyte blood types. This was unexpected, because tests for the large number of different, genetically controlled blood antigens in cattle had shown that identity of blood types in sibs is infrequent in cattle. Nonidentical twins should be as rarely alike in blood types as sibs, because sibs and such twins are genetically equivalent.

It had long been known that most pairs of cattle twins of different sex consist of a normal male and an intersexual sterile female partner called a "freemartin." This peculiar situation had been explained by Tandler and Keller (1911) and Lillie (1917). They showed that twin embryos in cattle are usually joined by anastomoses of their blood vessels and that the production of a freemartin is related to the joint circulation of the twins. They assumed that the sex hormones of the male twin may reach the female twin and result in its maldevelopment as a freemartin. Owen recognized that not only are hormones shared by twins with joint circulation, but that primordial blood-forming cells may become transported from one embryo to the other. When such cells establish themselves in the tissues of their host, a permanent blood mosaicism will have been initiated with both the host's own and his twin partner's cells participating in the formation of blood. This theory means that the identity of the blood types in the cattle twins is not caused by presence of one and the same genotype in both twin partners, but is due to the presence in both of a mixture of blood cells of two different genotypes. This was proven to be true. The proportion of the two kinds of circulating erythrocytes varies from one twin pair to the other, but is usually very similar in the two partners of a pair.

The mosaic nature of blood-forming cells in heterosexual cattle twins rests not solely on immunological evidence. Leukocyte cultures from male twins of heterosexual pairs showed cells of XY as well as XX chromosomal constitution (Dunn et al. 1966), and the same was seen in female twins of such pairs (Fechheimer, Herschler, and Gilmore 1963; Makino, Muramoto, and Ishikawa 1965). Surprisingly, Makino and his associates found that the XY/XX mosaicism of both kinds of twins extended to other than blood-forming types of tissues, of mesodermal and endodermal derivation: kidney, interstitial, gonadal, lung, and liver. These new observations reopen the question concerning the causation of the intersexual development of the freemartin. Instead of, or in addition to, an endocrine effect exerted by the male twin partner, it is possible that the wide ranging XY/XX mosaicism of freemartins may be decisive.

Blood mosaics of corresponding origin have also been found in sheep (Stormont, Weir, and Lane 1953), chickens (Billingham, Brent, and Medawar 1956) marmosets, a group of primitive primates (Benirschke and Brownhill 1962), and in man (Dunsford et al. 1953; review of later cases in Race and Sanger 1962). Human twin pairs of opposite sex are normal in their sexual development. Not only are anastomoses between their placental blood vessels rare, but, more important there is in general no interference in the development of the genital organs of either twin by the other. Nevertheless, the occasional anastomoses that do occur give an opportunity for exchange of blood-forming cells that may settle in the twin embryos. As examples of the resulting mosaicism, we may cite Miss Wa. and her twin brother, both of whom carry two populations of blood cells of A_1 and O type. The proportion of the two types is greatly different in these twins, Miss Wa. having 99-percent O and 1-percent A_1 cells, Mr. Wa. having only 14-percent O and 86-percent A_1 cells. It was possible to determine which of the cells were each twin's own and which originally came from the other partner. Miss Wa.'s salivary cells secreted the H substance which is related to O, and no A substance. Thus her constitution was O, and the A_1 cells were acquired. Reciprocally, her brother secreted the A substance in his saliva, showing that the A_1 cells in his blood were his own and the O cells acquired. His mosaicism for blood cells was also proven by an independent method, that of cytological study of his polymorph cells (Davidson, Fowler, and Smith 1958). In females these cells contain nuclei which often show a small protuberance known as a "drumstick." This is absent in polymorphs of males. Mr. Wa.'s polymorphs showed drumsticks with a frequency that was lower than in females and compatible with the proportion of O cells embryonically derived from his sister.

In another heterosexual pair of human twins, both partners had two red-cell populations, consisting of about 85-percent group O and 15-percent group A. Chromosomal study of 100 cells from blood cultures showed that the male twin had 70 cells with XX and 30 with XY constitutions, not very different from the female twin with her 78 XX and 22 XY cells (Uchida, Wang, and Ray 1964).

Transfer of primordial blood cells in utero may also occur between mother and fetus, as was recently shown to be probable by Kadowaki and others (1965). They describe a single-born male subject whose blood proved to contain a mixture of XY and XX cells. Several other tissues of the infant were studied and were apparently nonmosaic, being composed of XY cells only. The infant was ill with a severe immunological deficiency syndrome of unusual type that could be accounted for on the basis of an early intrauterine "graft" of maternal cells. Although other explanations for this mosaicism are not excluded, it seems likely that the XX cells were of maternal origin.

Triple blood mosaicism has been observed in one of the partners of a pair of dizygotic cattle twins (Stone, Friedman, and Fregin 1964). When first tested at the age of one month and again at three years, both twins had two populations of blood cells present in the same proportions, one constituting 90 percent of the cells ("Type I"), the other 10 percent ("Type II"). Five years later, one of the twins was no longer available but the other now possessed three types of blood cells. Each of the earlier Types I and II made up only 2 percent of the blood cells, and a new type of cells combining the antigens of Types I and II formed the remaining 96 percent ("Type III"). The authors interpret the origin of the mosaicism by assuming two successive events. The first is the "orthodox" process of exchange of some of the blood-forming cells between the twin fetuses. The second is a process of fusion between somatic cells including fusion of their nuclei, which has been proven to occur in tissue cultures of cells from mice and other mammals (Barski, Sorieul, and Cornefert 1960; Sorieul and Ephrussi 1961). Such fusion between cells belonging to the original two types in the blood-forming tissue of the twin would indeed account for its "somatic hybrid" blood cells. An alternative, however, cannot be excluded. Possibly the pregnancy that led to the birth of the cattle twins actually began as a triplet pregnancy. If so, the triple complement of blood-cell types could have been the result of the establishment of three initially genetically different cell populations in the triple mosaic, one of which happened to contain a combination of the antigens present in the two others. An argument against this interpretation is the fact that Type III cells had not been present in demonstrable numbers when the animal was young, although a triple origin would have demanded that all three types coexist from embryonic life onward. This argument does not weigh heavily, however, because striking changes with age in the proportion of two cell components in the usual kinds of blood mosaics in cattle as well as in man are well documented.

Germ Line Mosaics? A particularly intriguing possibility concerning transfer of cells between twin embryos has recently been recognized. It refers to the primordial germ cells. In higher vertebrates such as birds and mammals, as well as in other animals, the germ cells of the embryo do not originate in the developing gonads but migrate into them both by means of ameboid movement within the body tissues and by being carried to the gonads in the blood stream (Witschi 1948). In cases of anastomoses between the placental circulations of twins, an opportunity is thus provided for the settling of germ cells derived from one twin in the gonads of the other.

There are some breeding tests with blood-mosaic male cattle twins in which each animal produced offspring corresponding to a single genotype, that of one of the blood-type populations (Stone et al. 1960; Dunn et al. 1966). It is possible, however, that the number of such breeding tests did not suffice to settle the question whether germ cells of a different derivation were also present. Cytological evidence for the presence of two populations of germ cells within the gonads of single individuals of pairs of twins has been obtained in cattle (Ohno et al. 1962; Ohno and Gropp 1965) and in marmosets (Benirschke and Brownhill 1963). Ohno and his colleagues studied the chromosomal constitutions of spermatogonial cells in the testes of newborn male calves of heterosexual twin pairs and found not only the expected XY types but also numerous cells with two X-chromosomes and no Y. These cells were thus of typically female XX type. Presumably they had been derived from germ cells of the female co-twin.

It was not clear whether the XX cells in these testes were able to differentiate into spermatocytes and ultimately into sperm. A partial answer to this question came from the study of marmosets. Pregnancies in these animals usually consist of fraternal twins, and placental anastomoses between the twins are the rule. As a consequence of this situation, approximately one-half of adult marmosets are mosaic for XX and XY bone-marrow cells. The other half, with XX or XY cells only, is likely to be mosaic also, but this is not demonstrable by chromosome inspection: pure XX-cell populations being composed of the animal's own and descendants of her twin sister's embryonic cells and pure XY-cell populations similarly composed of two different XY subpopulations. When the kidney, skin, and testes of adult male marmosets with proven blood mosaicism were studied, only XY cells were found in the somatic tissues. The testes, however, in two of the animals contained XY as well as, apparently, XX cells. Such cells were not only of presumably spermatogonial type but were also spermatocytes undergoing first meiotic division. If confirmed, this is a most interesting contribution to the causal analysis of germ-cell differentiation. It might indicate that at least in a primate the differentiation of male germ cells is independent of their sex-chromosomal constitution and therefore solely determined by the nonger-minal tissue of the gonad in which they are located. It will be important to obtain information on the reciprocal situation, the possible existence of an admixture of XY cells in ovaries of mosaic XX mammals. Such information is more difficult to obtain and is not yet available.

A new discovery, such as that of gonadal mosaicism due to germ-cell migration from one twin to the other, often leads to insights in what seem to be only distantly related fields. A striking example of this is a recent study on the mule, the hybrid between horse and ass. Mules are nearly always sterile, a fact which seems to be related to the great differences in the chromosome complements of the parent species. These differences interfere with meiosis. There are, however, cases on record claiming that female mules have had offspring. Benirschke and Sullivan (1966), who feel that the evidence for fertility is not fully satisfactory, have shown, nevertheless, that ovarian development may lead to the formation of corpora lutea, which suggests the production of fertile eggs. To account for such eggs without assuming the improbable event of normal meiosis in mule oöcytes, Ohno (in Benirschke and Sullivan 1966) has suggested the following hypothetical course of events. A fertile mule may be a mosaic who had obtained some of its germ cells from a twin partner by way of anastomoses of embryonic circulation. The twin partner itself would have originated not from the hybrid combination between horse and donkey but from an independent mating of the mother of the hybrid with a male of her own species. Specifically, a mule derived from the mating of a horse mare to a donkey stallion might be twin to a horse fetus from another mating of the mare to a horse stallion (Fig. 28). Correspondingly, a hinny derived from the mating of a female donkey to a horse stallion might be twin to a donkey fetus from another mating of the same female to a donkey male. According to this speculative course of events, a fertile mule or hinny would essentially represent a fertile egg cell of the pure maternal species enclosed in the body of an infertile hybrid.

One may well foresee the discovery that human gonadal mosaicism may arise as a result of transfer of germ cells from one twin embryo to another. Perhaps some of the very rare cases where the blood-group genotype of a child seems to exclude the maternity of his proven mother will find their explanation in the fact that the child, though conceived by the mother, began as an egg cell derived from the mother's twin rather than from one of her own germ cells. Expressed as a

---

FIG. 28. A possible interpretation in terms of germ line mosaicism of the rare fertility of mules and hinnies. See text for explanation.

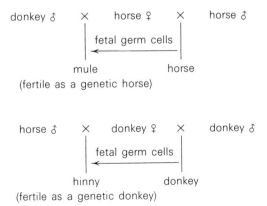

paradox, the genetic mother of such a child would not be the woman who bore him but his aunt if there was a female twin, or even his uncle if a fertile egg cell was able to differentiate from a male twin's germ cells.

It has been customary to designate mosaics composed of cells derived from two independent individuals as chimeras (named after the mythical monster that was compounded of a lion's head, a goat's body, and a serpent's tail). Chimeras were thus distinguished from individuals whose cells were all derived from the same single fertilization event but in whom subsequent chromosomal or genic changes had brought about the existence of more than one genetic cell type. The designation "mosaic" was reserved for such compounds. The distinction between the two types of compounds breaks down, however, if one considers the cases of double fertilizations in insects, mice, and man in which two sperm and one or two egg nuclei from the very beginning contribute their different genotypes to the developing embryo. Moreover, cytological proof of mosaicism usually does not provide evidence for its origin. Although mosaicism may be most easily understood in terms of abnormal distribution of chromosomes after an initially typical, single-egg-nucleus/single-sperm fertilization, in most cases only genetic evidence can exclude an origin from double-fertilization events. It seems appropriate therefore to use the term "mosaic" for any kind of genetic multiplicity in an individual, as is done in this essay, and to leave to specific inquiry the decision concerning what circumstances led to its origin.

Mosaicism and Antibody Formation. The types of mosaicism discussed thus far have in common the fact that they were the results of relatively rare and accidental events. Surprising and important results of recent studies show that, in addition, there are classes of mosaicism which are part of the normal constitution of organisms. One of these is the mosaicism of immunologically competent cells in individuals of various animal groups including birds and mammals. A single individual is able to produce a great variety of proteins, each of which acts as an antibody to a specific antigenic substance. How the antibody-producing cells, primarily the "plasma cells," are able to form so many different kinds of antibodies has been under discussion for many years. One theory assumes that the plasma cells of an individual are fundamentally alike and that the specificity of their products is secondarily impressed on the products when they are in actual contact with the different antigens. Another theory postulates that there is a great genetic diversity among the plasma cells which specifies a great variety of antibodies before there is any occasion of a confrontation with specific antigens. According to this theory, the introduction of an antigen results simply in a selection for growth stimulation of those cells which pre-adaptively happen to have acquired the genotype for appropriate antibody production (see Lederberg 1959). If one assumes the validity of this theory, the question arises how the genetic mosaicism of the plasma-cell population originated. Again, several alternative hypotheses have been proposed. One of these assumes that the genes of plasma cells which code for antibody proteins are unusually unstable so that different cells come to possess different alleles of such genes. Whether this high mutability of plasma cells is restricted to the antibody-controlling genes or whether it applies to the whole genome of these cells is left open. Another hypothesis assumes that the antibody-controlling genes are from the outset structurally peculiar, and that somatic recombination processes occur frequently and result in new types of structural genes (Smithies 1963). In detail it is assumed that the initial structural peculiarities of the genes consist in their having duplicated sections of nucleotide sequences. The initial form of this hypothesis assumed that the duplicated sections are inverted in relation to each other. The following discussion will proceed on the basis of this asumption, but other hypotheses involving somatic recombination have since been proposed.

The origin of inverted duplications of nucleotide sequences would have been due to "accidents" that happened to the genic material during the course of evolution. The results of these chance events then became established in the species or larger taxonomic group by means of natural selection. Inverted repeat sequences are known to be able to fold back on each other and thus provide opportunities for recombination not only between homologous strands of DNA double helices and between sister chromosome strands, but also between homologous sections within a single chromosome strand, such as a single double helix.

The result of intrastrand recombination is shown in Fig. 29. Schematically, a short section of a DNA molecule is represented by eight triplets of nucleotide pairs, 1 to 8, which code for amino acids a to h. Two of the triplets, 2 and 3, are present in duplicate but are inverted relative to each other so as to constitute also 6 and 7. As a result of recombination between the inverted repeat sections, the original DNA molecule becomes changed. The region lying between the repeats is now present in inverted sequence. This means that coding for the amino acids d and e exists no more, but that new codons may have originated for two new amino acids, i and j. It is easily seen that regardless

FIG. 29. Diagram of a section of a DNA gene consisting of 8 codons. *Top,* the triplets 2 and 3 are duplicated, in inverted sequence, as triplets 6 and 7. *Middle,* the duplicated triplets undergo pairing. As a result the section bends back on itself. Recombination occurs within triplets 3 and 6 between complementary strands. *Bottom,* the section of 8 codons after intragenic recombination. Triplets 1, 2, 3, 6, 7, and 8 are unchanged, but the original triplets 4 and 5 are replaced by two new triplets 4′ and 5′ as a result of inversion of the stretch between the duplications.

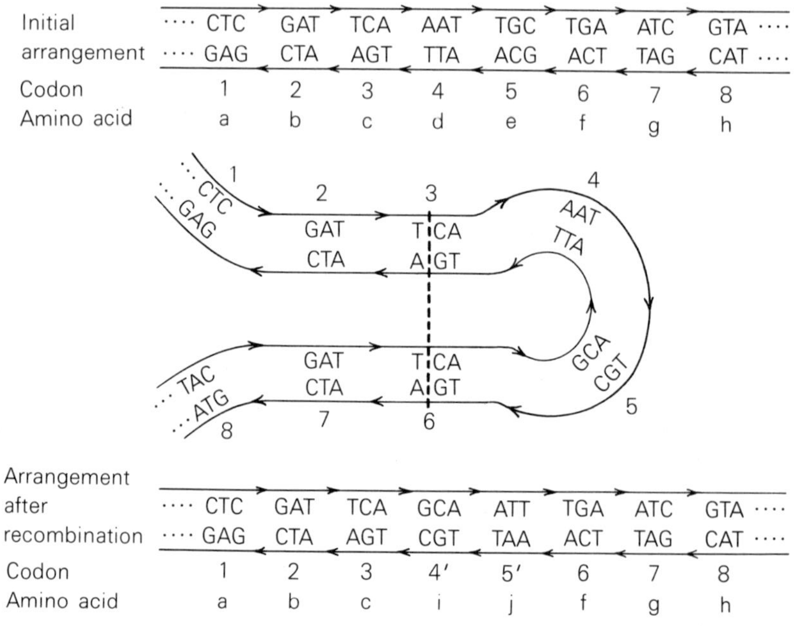

Initial	···· CTC	GAT	TCA	AAT	TGC	TGA	ATC	GTA ····
arrangement	···· GAG	CTA	AGT	TTA	ACG	ACT	TAG	CAT ····
Codon	1	2	3	4	5	6	7	8
Amino acid	a	b	c	d	e	f	g	h

Arrangement								
after	···· CTC	GAT	TCA	GCA	ATT	TGA	ATC	GTA ····
recombination	···· GAG	CTA	AGT	CGT	TAA	ACT	TAG	CAT ····
Codon	1	2	3	4′	5′	6	7	8
Amino acid	a	b	c	i	j	f	g	h

of the length of the region lying between the repeats, crossing-over within the paired repeat sections will lead to inversion of the whole section. Thus, if this section consisted of, for instance, ten triplets, their whole sequence would be inverted, thus possibly coding for ten amino acids different in kind and sequence from the original series.

Given the possibility of intragenic, intra-chromosome strand recombination, existence of one repeat within a gene can result in two different proteins, those of the original and those of the recombined type. If n different repeats should be present in the gene and intrastrand crossover results are combined in all possible ways, coding for 2^n different kinds of proteins would be provided in the assembly of antibody-producing cells.

The theory of somatic intrastrand crossing-over involving antibody-producing genes does not specify whether such crossing-over is restricted to the antibody-producing cells or whether it also occurs in other cells of the body. If it did it would not result in immunological effects. In Drosophila there is evidence for variable frequencies of somatic crossing-over in different tissues, and it may well be that natural selection has facilitated the occurrence of intrastrand recombination in the antibody-producing cells.

It is not yet possible to decide on the basis of experimental evidence between the mutation and the recombination theories of the assumed genetic mosaicism of plasma cells. Methods for reaching such a decision are available. It should soon become known whether different antibodies in a given species show only narrowly localized amino-acid differences or whether they differ in whole sequences of amino acids. The first finding would be in favor of the mutation theory of plasma-cell mosaicism, the second in favor of the theory of recombination by somatic crossing-over. It must be added that theories other than the two briefly treated here are also under discussion.

Functional Mosaicism. The evidence is increasing that differentiation in development is correlated with variable gene activity. In some cells and tissues certain genes are inactive and others active; in other cells and tissues inactivity and activity apply to other gene assortments. According to this concept, all multicellular organisms are mosaics in a functional sense even though most of them are uniform in having the same genetic endowment in all somatic cells.

The regular functional mosaicism of differentiation transcends the present discussion and will not be dealt with any further. There exist, however, two types of irregular functional mosaicism which should be described here. The first concerns a special type of sexual differentiation, the second the so-called variegated-position effects.

Mosaicism in Intersexes. The term *intersex* was introduced by Goldschmidt (1915) in order to designate a type of organism intermediate between a normal female and a normal male, and different from a gynander. As we have seen, gynanders are mosaics consisting of mixtures of genetically pure male and pure female parts. Intersexes, on the other hand, are genetically uniform individuals and their intermediate sexual appearance is the result of abnormal sexual development under the influence of unusual environmental conditions or of a special genotype. One may distinguish two general types of intersexes, regular and irregular. In regular intersexes the expression of sexual traits is rather constant. Different individuals are alike in having either certain whole parts characteristic of females and other whole parts characteristic of males or in regularly having some specific structures with intermediate and other specific structures with typical sex differentiation. Irregular intersexes, on the other hand, in different individuals show a wide range of variability in sexual expression for one and the same organ, a variability, moreover, that is expressed in mosaic fashion.

In man, testicular feminization with its differentiation of female breasts and female external and partially internal genitalia, but of male gonads, is an example of a regular type of intersex. Genetically this is a modified male. Another example is provided by intersexes in *Drosophila simulans*, which genetically are modified females (Sturtevant 1920). They possess various normal female structures, such as an ovipositor and spermathecae, and also male structures, such as lateral anal plates and claspers. A third example is the "doublesex" mutant in *Drosophila melanogaster*. Here both XX and XY individuals have simultaneously male and female systems of genital ducts and, in addition, sexually intermediate structures (Hildreth 1965). The forelegs of these intersexes are sexually intermediate structures. They are neither typically male nor female in terms of presence or absence of a sex comb. Instead, the arrangement of certain rows of chaetae is intermediate between those in normal males and normal females (Fig. 30). Moreover, the form of the chaetae is intermediate between typical bristles as in females and sex-comb teeth as in males.

As curious as these intersexes are, in principle their development does not differ from that of other regular types of differentiation. Different parts of their body are different in an orderly way, and the fact that both male and female structures are found is, in a way, similar to formation in the same individual of two different organs such as a brain and a limb. There are, however, kinds of intersexes whose phenotype varies in an irregular mosaic fashion from one individual to another. This was the type of intersexes which Goldschmidt encountered in the gypsy moth *Lymantria dispar* (review, Goldschmidt 1934). They appeared in the offspring of crosses between sexually normally differentiated geographic races, and Goldschmidt explained their appearance on the basis of his balance theory

---

FIG. 30. Basitarsi of normal and doublesex (dsx) XX and XY individuals of *Drosophila melanogaster*. The normal XX females have a distal transverse row of relatively few bristles, and the normal XY males have a longitudinally arranged sex comb with many teeth. The doublesex individuals have a row of relatively numerous heavy bristles arranged in an intermediate direction. Bristles of most transverse rows are represented by outlines of their sockets and bracts. (From Hildreth 1965.)

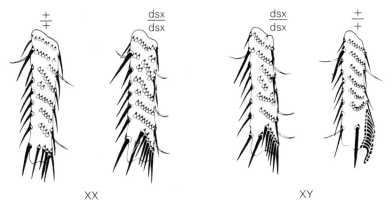

$$\frac{+}{+} \qquad \frac{dsx}{dsx} \qquad \frac{dsx}{dsx} \qquad \frac{+}{+}$$

XX   XY

of sex determination. According to this theory, both males and females possess genes for maleness (M) and femaleness (F). If it is assumed that the M-gene or genes are located in the X-chromosome, the two sexes would be distinguished by the dosage of these genes, two doses in the male moth and one in the female. If it is further assumed that the effect of the F-gene or genes is the same in both sexes, it follows that maleness results from 2M outweighing F, and femaleness from F outweighing 1M. Different geographic races are characterized by different "strengths" of the M and F determinants which, however, within each race balance each other in normal fashion. This is not so in hybrids between the races. In such hybrids combinations may occur in which neither M nor F overcomes its opposing determinant, and thus an intersex develops.

This formal genetic scheme alone does not account for the remarkable phenotypes of the Lymantria intersexes. Although some of their organs or parts seem to show a uniform, intermediate sexual morphology, in many respects certain of these intersexes are gynander-like mosaics. Their male and female structures are intermingled in different ways from one specimen to the next (Fig. 31). One could have postulated that the intersexes actually were gynanders if it were granted that an ex-

---

FIG. 31. *Lymantria dispar*. Above, normal female and male. *Below*, two diploid intersexes. (Redrawn from Goldschmidt 1934.)

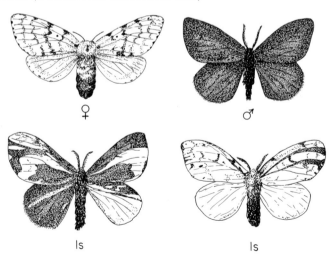

♀ ♂

Is Is

traordinary amount of repeated loss of sex chromosomes occurs during the ontogeny of geographical hybrids. Neither Goldschmidt nor anyone following him was prepared to make such a claim. It would have been impossible either to prove or disprove it on account of a lack of sex-linked marker genes and unfavorable cytological conditions. By general consensus it was accepted that the intersexes are diploid and of uniform sex-chromosomal constitution in all their somatic cells. Why then were they sexual mosaics? This question was answered by Goldschmidt in terms of a time law of intersexuality. The law stated that an intersex is an organism which begins development in terms of differentiation of one type of sex and which, after a "turning point," completes development in terms of the other sex. Parts which were determined before the turning point would continue their differentiation according to the early sex, while those determined after the turning point would develop according to the later sex. Finally, intermediate sexual differentiation would occur when the turning point coincided with a period of determination. Goldschmidt's doctrine was epitomized by his calling intersexes "mosaics in time," in contrast to gynanders which he called "mosaics in space," terms which do not fully withstand analytical scrutiny but which were the products of a period when Einstein's theory of relativity influenced everyone's thought.

Intersexes of irregular mosaic type were later obtained by Seiler in the moth Solenobia. They occurred in diploids as well as in triploids, the latter produced by crosses of tetraploid females and diploid males (review, Seiler 1965). In a series of most detailed and careful investigations it was shown that wherever cells were morphologically different in the two normal sexes, they would in intersexes differentiate into a mosaic of pure male and pure female parts (Fig. 32). The relative extent of these sexually different parts varies from one individual to the next, and the same individual may have both large and small patches of each sex in the same structure. Seiler felt unable to reconcile this variability with the time law. Instead, he concluded that an intersexual genotype produces an unstable equilibrium of the two normal sexual tendencies which changes to a stable male or a stable female tendency in different cells. In some cells this change occurs early in development and subsequently results in large patches of either sex; in other cells it occurs late and results in small patches.

In *Drosophila melanogaster* the morphology and the sex apparatus of "triploid" intersexes (2X3A) was studied by Dobzhansky and Bridges (1928) and Dobzhansky (1930). They found that, as in Lymantria, many of these intersexes are mixtures in various proportions of nearly typical male and female structures. The interpretation given by these authors was in terms of Goldschmidt's time law, although, somewhat later, Bridges (1932) pointed out that "the idea of a single turning point for the individual as a whole must be looked upon as an idealization and not as a fact." A similar conclusion was reached by Kerkis (1934), and, still later, Bridges (1939) withdrew completely his support of Goldschmidt's interpretations.

The irregular mosaicism of triploid intersexes in Drosophila is often strikingly apparent in the coloration of the abdomen

---

FIG. 32. Parts of intersexes of *Solenobia triquetrella*. (A) Pupal cuticle of the 7th segment with a mosaic of pure female and male areas. *Left middle,* numerous small thorns typical of male; *right,* fewer, larger thorns typical of female. The cuticle is also a sexual mosaic for pigmentation, being more darkly pigmented in the female than in the male region. Since the female abdomen grows more rapidly than that of the male, folds form in the mosaic at the border of the two parts. (B) Proximal part of the gonoduct. Mosaic of oviduct epithelium, which is flat, and sperm-duct epithelium, which consists of columnar cells. (A from Seiler 1949; B from Seiler 1958.)

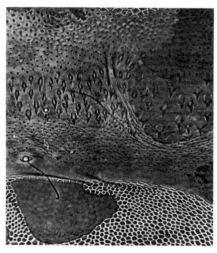

A   B

(Stern 1966). Small or large unpigmented areas typical for female-ness may occur within pigmented areas typical for maleness; whole dorsal segments may be female on one side of the midline and male on the other side, or even a whole abdomen may have the coloration of one sex on its left half and that of the other sex on its right (Fig. 33). The similarity of such mosaicism to proven gynandrism is striking. It led to the use of genic markers that would give unequivocal evidence for or against the as-sumption that chromosome elimination or special types of somatic crossing-over account for the mosaicism of intersexes. Not unexpectedly, the evidence proved to be against chromo-somal diversity between the male and female areas.

It cannot be said that the developmental genetics of mosaic intersexuality is understood. In the 2X3A intersexes of Dro-sophila the balance between the female-determining genes in the X-chromosomes and the male genes in the autosomes is in-termediate between that of 2X2A females and 1X2A males. It is not clear, however, why this results in alternative sex dif-ferentiation instead of intermediate development. A capacity for intermediate development does exist, as shown by the sex-comb region on the foreleg of the diploid doublesex mutant pictured earlier (Fig. 30). Why then does a triploid intersex in the majority of flies form either typical male teeth or typical female bristles? It may be hypothesized that the diploid intersex

FIG. 33. Mosaic pigmentation of the abdomen of *Drosophila melanogaster*. (*A*). 2X3A intersex. Terminal tergite and right half of other tergites, male type; left half of all except terminal tergite, female type. (*B*) 2X/1X gynander, redrawn from Morgan and Bridges (1919). Left, female; right, male. In both *A* and *B* note the greater width in female parts of the pigmented band on tergites 1–2, 3, and 4. (From Stern 1966)

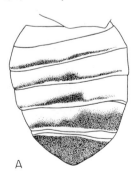

A   B

mutants influence differentiation during secondary stages, while the triploids determine sex at a more fundamental level—but how meaningful is such a formulation?

Variegated Position Effects. A male Drosophila whose single X-chromosome carries the normal allele at the white locus has red eyes. A female who carries the normal allele in one X-chromosome and the white allele in the other is also red-eyed, red being dominant over white. If, however, as a result of chromosome breakage and reunion, the red allele has been removed from its typical position in the so-called euchromatic part of the X to one near the "heterochromatin" (see Brown 1966) of either the X-chromosome or one of the autosomes, it is not able to function normally. The eyes of a male whose red allele has been permanently placed into or near heterochromatin are irregular mosaics of red and white (or at most light-colored) patches, and so are the eyes of heterozygous red/white females whose red allele is in the changed position (Fig. 34).

When such "variegated" flies were first discovered, they were designated "eversporting displacements," implying that the mosaic phenotype was evidence of highly frequent somatic mutation (Muller 1930). Later, various facts made it much more

FIG. 34. Variegated position effect in the strain "mottled white 258-18" of *Drosophila melanogaster.* (A) An eye with areas of red (wild type) and cherry pigmentation on a background of cream. (B) An eye with areas of cream and cherry pigmentation on a background of red. Mottled white 258-18 involves a translocation between the X-chromosome, broken close to the white locus, and chromosome 4. (After Demerec and Slizynska 1937.)

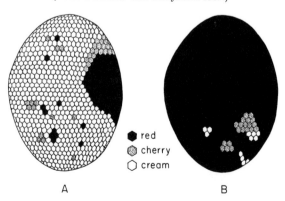

● red
◉ cherry
○ cream

A   B

likely that the molecular structure of the gene is not altered by its new environment in the chromosomes but that its action is. Variegated position effects leading to irregular mosaicism are known for many Drosophila genes, and include position changes from eu- to heterochromatin, as well as in the reverse direction.

More recently position effects have also been observed in the mouse, where they exhibit features similar to those in Drosophila, but they also differ in important ways. The known variegated position effects in the mouse affect the functioning of autosomal genes translocated to an X-chromosome (L. B. Russell and Bangham 1959, 1961; Cattanach 1961). For example, irregular mosaicism is exhibited by females, one of whose two X-chromosomes carries the translocated autosomal allele for the dominant fur color of the wild mouse and one of whose unchanged autosomes carries the recessive allele for brown fur: both wild-type and brown patches are present. This is as in Drosophila, but, unlike Drosophila, mouse males whose single X-chromosome has the translocated wild allele, and one of whose autosomes carries brown, are not mosaic: they show only the dominant wild-type fur. The same is true in mouse females of the exceptional XO constitution, given the same genetic setup as in males except for the absence of a Y-chromosome. These findings indicated that in the mouse there is a specific connection between singleness or doubleness of the X-chromosome and position-effect mosaicism. The nature of this connection soon became recognized as an expression of very general properties of the mammalian X-chromosomes. It will be described in the following section.

Functional Mosaicism of the X-chromosomes in Non-human Female Mammals. Intersexes and position effects are rare "freaks of Nature," and it is perhaps not too surprising that they are peculiar in more than one way, including functional mosaicism. It did, however, come as a complete surprise when it was discovered that all normal females, in man and other mammals, are functionally irregular mosaics. Three different lines of evidence converged toward this realization, one line involving cytological studies and two lines involving genetic studies. The first group of investigations began with Barr and Bertram's discovery of the "sex chromation," the small, stainable body in the nuclei of cells of female cats and its absence in cells of males (1949; Barr, Bertram, and Lindsay 1950). After initial suggestions that two X-chromosomes were necessary to form the Barr body, it was shown by Ohno, Kaplan, and Kinosita (1959) that in the rat the two X-chromosomes of the female do not behave identically. One condenses during prophase at the same rate as the autosomes, while the other shows positive heteropycnosis, that is, it is more condensed than the other chromosomes. The same applies to the X-chromosomes of somatic nuclei of human females (Ohno and Makino 1961). These facts led to the conclusion that the Barr Body is formed by only one chromosome, the heteropycnotic X-chromosome. Simultaneously, observations on cells of the Chinese hamster showed that during mitosis the two X-chromosomes of female cells differ in appearance (Yerganian 1959). One of them possesses a secondary constriction indicative of nucleolar development, while the other X-chromosome lacks the constriction. The single X-chromosome of the male is always constricted. These differences in chromosome behavior were studied further by Taylor (1960) who used autoradiography following the incorporation of tritiated thymidine. It became apparent that in male cells replication of DNA in the shorter arm of the X-chromosome occurs during the first half of the period of DNA synthesis. In female cells the short arm of one X-chromosome also replicates during the first half, but that of the other X-chromosome is delayed in replication until the second half. A striking asynchronous replication of the X-chromosomes has also been described in human females (Fig. 35; German 1964).

It was at this stage of the cytological inquiries that Liane B. Russell (1961) and B. M. Cattanach (1961) described the genetic position effects in the mouse and recognized their dependence on the presence of more than one X-chromosome in the cell. Russell explained the mosaic phenotype of translocation heterozyzotes as due to "the action of the wild-type allele . . . [having been] made 'uncertain'." As a result, this allele exerts its dominance in some cells only and not in other cells. In the latter the recessive allele becomes phenotypically expressed. Russell considered her own genetic findings together with those of the cytologists and concluded that "when two X's are present, heterochromatic action of one is possible and produces the

Fɪɢ. 35. Duplicate photographs of a radioactive lymphocyte in metaphase from blood of a normal human female, showing late replication of one of the two X-chromosomes. During the final 6 hours the cell had access in its medium to tritiated thymidine, and during the final 3 hours to colcimid. *Upper left*, orcein-stained; *upper right*, orcein-stained and autoradiographed indicating regions of late DNA synthesis ("end of S" period). *Below*, the karyotype of the cell. Rows 1, 3, and 5, chromosomes arranged from the metaphase figure at upper left. Rows 2, 4, and 6, the same chromosomes from the metaphase figure at upper right. The chromosomes that can be recognized by morphology are 1, 2, 3, 16, 17, and 18. Autoradiography adds further by delineating one of the X-chromosomes by its late replication (other chromosomes delineated by autoradiography are numbers 4, 5, and 13). The remaining chromosomes are, in this cell, not delineated by autoradiography and are mounted in groups only. (From Dr. James German, Cornell University Medical College).

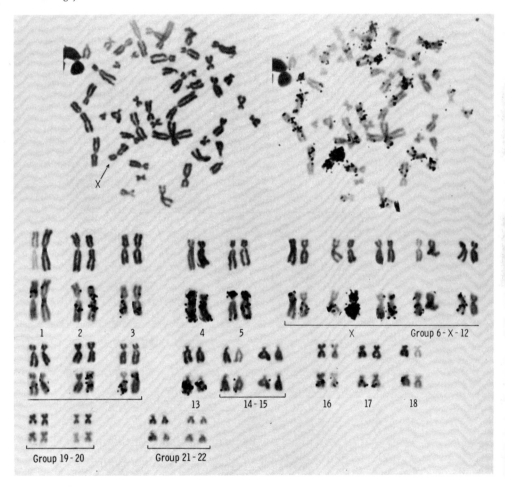

expression of the position effect . . . When, however . . . the rearranged X [is the only X present, then it] must fulfill the nonheterochromatic functions and will not be able to produce the position effect." These interpretations raised new questions. What determines which one of the two X-chromosomes of the female becomes heterochromatic in a given cell? When is such determination exercised? And why is the existence of the cytological differences between the two X-chromosomes of the female a normal phenomenon, when the genetic consequences assigned to these differences are only rare and abnormal situations leading to position effects?

Answers to these questions were proposed by Mary Lyon in a communication to *Nature* of less than one page (1961). Lyon started with two given phenomena. First, the normal phenotype and the fertility of the rare XO mice (Welshons and L. B. Russell 1959) showed that activity of a single X-chromosome suffices for normal female development. Second, all X-linked mutants affecting fur color known in the mouse at that time expressed themselves in mosaic fashion in heterozygous females. Some parts of the body surface exhibited wild-type and other mutant coloration. Based on these two facts, the hypothesis proposed by Lyon states that genetic inactivation of one X-chromosome in female cells is a normal process, randomly affects one X-chromosome in some cells and the other in other cells, occurs early in embryonic development, and is irreversible within any one cell lineage. These suggestions accounted for the mosaicism of females heterozygous for ordinary X-linked mutants as well as for the variegated position effects exhibited by autosomal genes translocated to an X-chromosome. One had only to assume that the inactivating nature of the X-chromosomal part affects also autosomal genes which have been transferred to it. No more discrepancy was left between the regular cytological behavior of the X-chromosomes and the behavior of genes located in or translocated to one of them.

Lyon's first communication on inactivation of one of the two X-chromosomes in females was mainly restricted to the interpretation of genetic data from mice. Later she extended her hypothesis explicitly to cover gene action in the X-chromosome of man and mammals in general (1962). Lyon has called her hypothesis that of the "inactive X-chromosome," while L. B.

Russell has preferred the designation "single-active-X" hypothesis. This emphasis on activity of one X-chromosome rather than on inactivity of its homologue takes account of the observations that even in cells with more than two X-chromosomes, only one is euchromatic and all the others are heterochromatic.[1]

Dosage Compensation. It may appear strange that evolution of the mammals involved a mechanism which would make useless one of the two X-chromosomes in the cells of normal females. Inactivation of one of the X-chromosomes is, however, one of the possible solutions to a problem that confronts all species in which the X-chromosomes occur twice in one sex and only once in the other. According to the balance theory of gene-dependent development, a certain dosage of all genes is required for harmonious interaction. In the case of autosomal genes, this dosage usually consists of two alleles for each gene. If, on account of a deficiency for an autosomal locus in one chromosome, only a single normal allele is present, or if in a heterozygote only one allele produces a normally effective protein, then only one-half of the normal gene product may form. Consequently, metabolism and development are liable to be disturbed.

Selection pressure, therefore, would maintain the diploid condition and activity of two alleles at each locus. All things being equal, this should be true not only for autosomal but also for sex-chromosomal genes. For these, however, all things are not equal. Unless X-linked genes have alleles in the Y-chromosome—and that is the case only rarely—one half of the individuals of each species have only single doses of them and the other half two doses. It would be expected that this situation would lead to different expressions of X-linked genes in the two sexes and that the two phenotypes would have unequal selective values.

In Drosophila most X-linked genes are expressed alike in males and females. Originally this was taken for granted, until it was recognized that inequality rather than equality in pheno-

1. Since the present account was written, Grüneberg has subjected the single-active-X hypothesis to a searching critical evaluation which led him to a rejection of it (1967). On the other side, new facts supporting the hypothesis have also come to light.

type should have been expected and that the observed equality implied the existence of a mechanism compensating for the 1 dose–2 doses difference (Stern 1929, 1960a; Muller, League, and Offermann 1931; Muller 1932, 1950; Cock 1964; Muller and Kaplan 1966). The mechanism in Drosophila is made up of complex genetic modifiers.

An equality in expression of certain sex-linked genes in males and females is also characteristic of man and mice, and it had been pointed out that this implied the existence of a system of dosage compensation in mammals (Stern 1960a). The specific nature of such a system had to be left to further inquiry. Within a year it turned out to be the process of inactivation of one X-chromosome in the female. In Drosophila dosage compensation seems to be accomplished by the male imitating the female—the single X-chromosome in the male presumably acting as strongly as the two X-chromosomes together in the female (Offermann 1936; Dobzhansky 1957; Mukherjee and Beermann 1965; Mukherjee 1966; Muller and Kaplan 1966). Mammals make the opposite choice. Instead of the males aspiring to equality with the females, the latter, by inactivation of one of their X-chromosomes, succeed in becoming equal to the males.

Inactivation of one X-chromosome is not the only mechanism of dosage compensation in mammalian females. Even when inactivation occurs, it seems to vary in significance. In human females its consequences are less than those of actual loss of an X-chromosome. This is shown by the fact that XO women are sterile and affected with a variety of unusual traits (Turner's syndrome). In mice the consequences of loss of an X-chromosome are hardly greater than those of inactivation, because XO mice are fertile and phenotypically indistinguishable from XX mice. The XO mice, however, produce fewer offspring than XX individuals. In some other mammals, the need for two X-chromosomes in females has been abolished completely. The male meadow vole *Microtus oregoni* has only a single X-chromosome in the somatic cells of both sexes, the male cells being XY, the female cells XO (Ohno, Jainchill, and Stenius 1963). The male germ line has no X-chromosome and is—rather unbelievably!—OY; the female germ line is XX. This complex situation seems to be the result of a sequence of unusual processes: formation of O and Y sperm, production of XO female and XY male zygotes, duplication of the X-chromosome in the female germ line, and elimination of the X-chromosome in the male germ line.

Still other types of mosaicism have been observed in three species of marsupials (Hayman and Martin 1965). Both males and females are alike in being XO in the cells of their liver and spleen as well as in leukocytes. The male germ line is XY and the female germ line (of the two species available for oögonial study) is XX. Here it appears that dosage compensation has led to elimination of one X-chromosome in the female soma and of the Y-chromosome in the male soma resulting in somatic identity of the two sexes. Additional surprises, however, seem in store. Culture lines derived from skin tissue of each of the two sexes of one of these marsupial species were found to be XY and XX respectively (Jackson and Ellem, cited in Hayman and Martin 1965).

The normal mosaicism of voles and marsupials belongs to a class which has been termed gonosomic, since gonad and soma of the individuals have different constitutions. Peculiar types of gonosomic mosaicism have long been known in nematodes of the genus Ascaris, the coccid Icerya, and the dipteran families of Sciaridae and Cecydomyidae (Beermann 1956, review). The evolutionary and developmental significance of this invertebrate mosaicism is not well understood. Dosage compensation seems at most a minor aspect of its existence.

Functional Mosaicism of the X-chromosomes in Human Females. Evidence bearing on the single-active-X hypothesis has come forth from a variety of studies. Only a few of these, particularly those on the human species, will be cited here. A survey of the literature on the expression of X-linked genes in women led Lyon (1962) to the conclusion that her hypothesis stood up to test so far but that more evidence was needed. She found that women heterozygous for so-called recessive genes in the X-chromosome often show an occasional manifestation of the recessive allele, sometimes in a mosaiclike fashion. This was to be expected if inactivation of one or the other X-chromosome occurred. Striking examples of such mosaicism had been reported for the sex-linked condition called anhidrotic ectodermal dysplasia, which results in nearly complete absence of sweat glands and absence of teeth in males carrying the abnormal gene. Heterozygous females are irregularly mosaic for normal and abnormal skin areas and their gums may show regions of normal teeth formation next to empty or defective sites (Fig. 36). Other traits such as hemophilia A and B that have usually been regarded as recessive actually show a wide range of expression in heterozygotes. The amount of clotting factors varies from less than 20 percent to 100 percent of normal, and it has been estimated that 2 percent of females heterozygous for hemophilia A may have clotting-factor levels low enough to result in clinical hemophilia (McKusick 1962, review). This variability may be interpreted in terms of variable proportions in different women of the two alternative functional genotypes of the cells concerned with elaboration of clotting substances. When the X-chromosome which carries the normal allele is

FIG. 36. Three generations of women heterozygous for presumably X-linked anhidrotic dysplasia. Stippled areas indicate regions of increased skin resistance to an electric current due to absence or low frequency of sweat glands. The two women of generation III are identical twins. Note the striking differences in the distribution of the low-resistance areas. (From Kline, Sidbury, and Richter: *J. Pediat.* *55*: 355–366 (1959), The C. V. Mosby Company, St. Louis.)

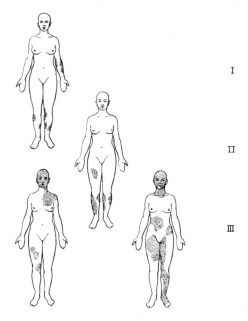

I

II

III

active in a certain minimum proportion of cells no heterozygous effect would be observed, but when this minimum is not reached, because the chromosome with the normal allele is inactivated in too many cells, mild or even severe symptoms would ensue. In this connection a chromosomally mosaic female described more recently is of interest (Fig. 37; Gilchrist, Hammond, and Melnyk 1965). She was the daughter of a non-hemophilic father. The mother was heterozygous for hemophilia A; her son was hemophilic and she herself had a low level of anti-hemophilic globulin. Normally, daughters of a non-hemophilic man are not bleeders. This was indeed true for two daughters in this family, but not for the proposita, the firstborn daughter. Her level of the anti-clotting factor was approximately as low as that of her brother, a typical bleeder, and accordingly she also was a severe clinical example of the disease. Cytological analysis of her leukocytes as well as of fibroblasts from a skin culture showed her to be a mosaic consisting of between 70 and 80 percent of cells with typical normal XX constitution and between 20 and 30 percent of cells with only one X-chromosome. It would seem suggestive that the hemizygous cells had lost the X-chromosome with the normal allele for hemophilia A and that the bleeding state of the patient was due to this component of her mosaic makeup. In itself, however, the chromosomal mosaicism is not sufficient to account for the low amount of anti-hemophilic globulin, because only an insufficient percentage of her cells were hemizygous. On the other hand, if inactivation of one of the X-chro-

FIG. 37. Pedigree of a family with an XO/XX daughter afflicted with hemophilia A. She is the oldest of four sibs. The level of anti-hemophilic globulin, stated in units of assay, is given below each individual. Normal levels vary between 52 and 133 units. Solid symbols: hemophilia; shaded: low assay values but not clinically abnormal. (After the data of Gilchrist, Hammond, and Melnyk 1965.)

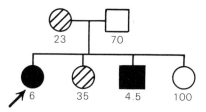

mosomes in her XX cells would by chance have affected the chromosome with the normal allele considerably more frequently than that with the abnormal allele, then the additional presence of 20-30 percent hemizygous hemophilic cells might have been decisive in shifting the phenotype to full hemophilia.

It is seen that the single-active-X hypothesis is flexible enough to be relevant to a variety of observations. Direct support for it comes from specific studies inspired by the hypothesis. Among these, the analyses of women heterozygous for alleles of a gene responsible for presence, absence, or specific variants of the enzyme glucose-6-phosphate dehydrogenase are of outstanding importance. Pedigree studies had shown that this gene is located in the X-chromosome. Females who are heterozygous for the allele responsible for G-6-PD deficiency show a remarkable spectrum of enzyme levels. Many women have approximately 50 percent of the normal amount of the enzyme but not infrequently they may be grossly deficient for it or be entirely normal. Beutler, Yeh, and Fairbanks (1962) demonstrated that women with intermediate enzyme levels possess two populations of red blood cells, one with normal activity, the other deficient. This is indeed what was predicted by the single-active-X hypothesis. Further evidence for two populations of red blood cells in females heterozygous for G-6-PD deficiency came from spectrophotometric examinations of individual blood corpuscles (Dreyfus, Bessis, and Kaplan 1964).

There remained the possibility that the heterozygotes are not functional mosaics in terms of the hypothesis, but that their cells vary for some other reason in such a fashion that a threshold superimposed on continuous variation divides these cells into two phenotypically alternative groups. This was disproven by the decisive study of Davidson, Nitowsky and Childs (1963), who obtained tissue cultures derived from single skin cells ("clones") of (1) women heterozygous for the enzyme deficiency, as well as (2) women heterozygous for two variants of the enzyme, A and B, which can be distinguished by electrophoretic study, A being a faster moving protein than B. The first group of women were Caucasians from Sardinia where the G-6-PD deficiency is relatively frequent. Regardless of whether cells from original skin cultures or from derived cloned cultures were tested, cells from hemizygous normal males and homozygous normal females showed an activity of the enzyme exceeding 5.2 units of a certain assay, while cells from hemizygous-deficient males and homozygous-deficient females tested at below 3.4 units. This was different in cultures from two heterozygous women. The original culture of one of them gave an activity of 6.3, a value which is close to the lower limit of normal, the other gave an activity of 2.4, well within the deficient range. When, however, cloned cultures were assayed, three strains from each woman had fully normal activity and one strain from each had typically deficient activity.

The second group consisted of six American Negro women who were heterozygous for the electrophoretic variants G-6-PD A and B. (Both variants are frequent in Negroes, but only B is known in whites.) The heterozygosity of the women was deduced from the existence of both A and B bands after starch-gel electrophoresis of extracts of cells from cultures prior to cloning. Subsequently, 54 clones of cultured skin cells of the six women were tested for presence or absence of the fast A and the slow B bands. There were 24 clones which belonged to type A and 30 which belonged to type B. Not a single clone was both A and B (Table 2). While one woman happened to yield only A and two women only B clones, the clones of three

---

FIG. 38. Electrophoretic pattern of G-6-PD properties of cell cultures from a woman who possesses the AB phenotype. Samples were started at the top of the figure. Left edge, from a culture prior to cloning showing two bands. Upper, slow band: B; lower, fast band: A. Clonal cultures 1-9: six cultures showing the B, three showing the A band only. (Redrawn from Davidson, Nitowsky, and Childs 1963.)

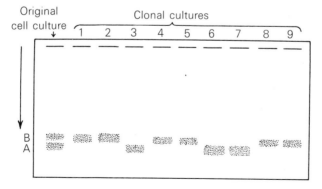

women included both types. Figure 38 shows the original double-banded electrophoretic picture of one of the last listed women, Mrs. De., and of nine of her cloned cultures, three of type A and six of type B. In summary, the evidence for irregular functional mosaicism of the X-linked G-6-PD alleles in human females is very impressive.

The large sizes of the two alternative areas in coat-color mosaics in mice, along with the early appearance of sex chromatin in the embryos of various mammals including man, suggested to Lyon that the onset of inactivation of one or the other X-chromosome occurs at a stage when only rather few cells have been formed in the embryo. It was unexpected, therefore, when Linder and Gartler (1965a) reported that most small tissue biopsies from eleven women heterozygous for the A and B variants of G-6-PD contained mixtures of both substances in various proportions. Several explanations are available to account for the fine intermingling of the two functionally different cell types exhibiting the activity of either the A or B allele. For instance, early inactivation of one X-chromosome in different cells may be followed by cell mixing instead of contiguous location of cells descended from a common ancestral cell. Or, inactivation of a specific X-chromosome may not always be fixed immediately, and a period of alternative inactivations among descendants of a cell may exist. Moreover, the large size of the patches in X-linked mosaics involving fur coloration in female mice or in ectodermal displasia in woman may not reflect completely alternative functional genotypes, but only the presence of a majority of cells with the gene activity

TABLE 2. Numbers of clones of cultured skin cells from women who possess the AB electrophoretic G-6-PD phenotype, as determined by starch-gel electrophoresis.

Type	Mrs. Bi.	Mrs. Mi.	Mrs. Wi.	Mrs. Bo.	Mrs. De.	Mrs. Ha.	Total
A	8	1	0	8	7	0	24
B	2	8	8	0	7	5	30
AB	0	0	0	0	0	0	0
Total	10	9	8	8	14	5	54

Source: Davidson, Nitowsky, and Childs 1963, slightly changed.

responsible for the phenotype of the patch. These unsolved questions do not detract from the value of the single-active-X hypothesis. Its validity is even strengthened by one other aspect of the distribution of the A and B variants of G-6-PD discovered by Linder and Gartler (1965b). They found that 28 out of 29 tumors of the uterus that had developed in women heterozygous for A and B were pure in having either an A or a B variant of the G-6-PD enzyme. (The exceptional twenty-ninth tumor may have included a relatively large admixture of non-tumor cells.) The authors point out that this is consistent with, but does not prove, the concept that these tumors arose from single cells. Reciprocally, one may say that if these tumors arose from single cells, their G-6-PD production is consistent with the single-active-X hypothesis.

The hypothesis has also been applied to a very serious pathological condition, the X-linked Duchenne type of muscular dystrophy. Women heterozygous for the abnormal gene are usually normal but often exhibit some minor defects associated with the trait and, very rarely, the clinical features of the disease. Biopsies of muscle tissue of presumed carrier females have revealed a mosaic of normal and abnormal muscle fibers, the latter with the appearance characteristic of muscular dystrophy in affected males. The distribution of the two types of fibers was such as to suggest two separate populations (Pearson, Fowler, and Wright 1963). It was later stressed that intermediate fiber types also occur, and that this is compatible with the observation that muscle fibers are produced by multinucleate cells formed by fusion of uninucleate cells (Emery 1964). Given this multinucleate origin of muscle fibers, one would expect various mixtures of nuclei with either the normal or the abnormal allele being active. Thus, one would expect some fibers with all or nearly all nuclei of one kind, others with more or less equal mixtures of both kinds, and still others with all or nearly all nuclei of the other kind. This reinterpretation of the original findings is still based on the validity of the single-active-X hypothesis.

An ingenious method of testing the hypothesis in a general way was suggested by Vandenberg, McKusick, and McKusick (1962). They reasoned that the two members of identical female-twin pairs should be less similar to each other than those of identical male-twin pairs. Identical male twins share the same X-chromosome but identical female twins, being irregular functional mosaics for their two X-chromosomes, should vary in the proportions and distributions of their two cell populations. It was first reported that a study of the intra-pair variability of many traits did indeed show a higher variance in female than in male identical twins. Later, errors were recognized by the authors and no excess variability interpretable as support for the single-active-X hypothesis was found to exist (McKusick 1962). Nevertheless, further study of twins along these lines may still yield valuable results.

It would be desirable to correlate the alternative inactivation of pairs of X-linked alleles directly with the alternative heterochromatization or differential labeling of the X-chromosomes. This has not yet been possible in man. Still lacking is even the more limited demonstration which should show that heterochromatization or differential labeling applies to one specific X-chromosome in some cells of females and to the other in other cells. There are some studies of women who have one structurally normal X-chromosome and one morphologically abnormal X-chromosome (Giannelli 1963; Lindsten 1963; Muldal et al. 1963; London et al. 1964). In all the cells of these persons the abnormal X-chromosome was late labeling. This at first seems contrary to expectation from the viewpoint of Lyon's random inactivation hypothesis, but the disagreement does not weigh heavily. For one thing, it could be assumed that the morphological and genetic abnormality of one of the chromosomes determines its late labeling and that the X-normal/X-abnormal situation does not justify arguments against the interpretation of the cytology of typical, X-normal/X-normal women. For another, it has been pointed out that cells in which only the abnormal X-chromosome is active are deficient for genetic material and are likely not to survive (Gartler and Sparkes 1963). Thus, perhaps, alternative inactivation actually did take place in the cells of the embryos which developed into the unusual females, with only those cells surviving in which the normal X-chromosome had remained active.

A related but in a way contrary phenomenon has been discovered in mice heterozygous for a normal X-chromosome and an X-chromosome that had undergone a reciprocal translocation with an autosome (Lyon et al. 1964). Here the genetic evidence indicates that it is the normal X-chromosome which is inactivated exclusively, while the translocation X-chromosome is the active one. The reason for this state of affairs is not clear, but two alternative hypotheses are discussed by the authors. One postulates actual random inactivation but death of the cells in which the translocation X is inactive, because of the associated inactivation of translocated autosomal genes. The other hypothesis considers some kind of possible interference with the normal process of inactivation due to the existence of the translocation.

If cytological proof for alternative inactivation of the X-chromosomes of normal women is still lacking, equivalent studies in mule and mouse are available. The X-chromosomes of female mules are of different kind, one being derived from the horse, the other from the donkey. Morphologically the two X-chromosomes are distinguishable by their different size and the position of the kinetochore. Mukherjee and Sinha (1964) obtained cultures of leukocytes from a female mule, treated them with tritiated thymidine, and by means of autoradiography studied the replication pattern of their chromosomes. Out of a larger number of cells at metaphase, 33 showed one chromosome that was heavily labeled in comparison with all others. In 16 of these 33 cells the heavily labeled chromosome was identified as the X-chromosome from the horse parent, and in 17 cells it was identified as the X-chromosome from the donkey parent. If this identification is accepted, the findings on the existence of two cell populations with late replication of either one or the other of the X-chromosomes constitute strong support for the single-active-X hypothesis. It may be noted, however, that the authors call their observations "preliminary" and that recognition of the X-chromosomes does not always seem to be unequivocal.

A very similar autoradiographic study has been made on female mice having one normal X-chromosome, X^n, and one much longer than the normal due to insertion into it of an autosomal section, X^t (Evans et al. 1965). The X^t-chromosome was clearly identifiable, being the longest chromosome in the complement, but the X^n-chromosome could not be distinguished from certain other chromosomes. Among over 2,000 labeled cells from bone marrow, spleen, and thymus of a mouse treated in vivo with tritiated thymidine, there were 180 that had a single late-replicating chromosome. In 92 cells this chromosome was X^t and in 88 cells it was presumably the X^n-chromosome. It must be added that in addition to these cells, which strikingly fit the single-active-X hypothesis, 7 cells were observed in which both the X^t- and the presumptive X^n-chromosomes were late replicating. Furthermore, Evans and his colleagues point out that the existence of a fourth class in which neither X is late replicating cannot be excluded. However, the argument which links late replication with genic inactivity is comfortably elastic so as to permit some leeway in the period of replication of chromosomes without embarrassing the basic reasoning.

There is, finally, one study in which cytological and genetic data bearing on the single-active-X hypothesis were obtained simultaneously. It made use of the same autosome-X-chromosome translocation in the mouse which had served for the labeling experiments just discussed. Ohno and Cattanach (1962) investigated the somatic prophases of skin cells and found, as expected from earlier work, that one chromosome in each cell was more condensed than the rest of the chromosomes (no autoradiography was employed). In some cells the condensed chromosome was of a size corresponding to that of the normal X^n, in others it was larger than X^n and corresponded to the translocation chromosome X^t. These cytological alternatives were correlated with a genetic alternative. The chromosome X^t carries the translocated dominant wild-type allele C for fur coloration, and the females heterozygous for X^t bore the recessive allele c for albinism on the corresponding autosome. The translocation chromosome is responsible for variegated position effect, so that inactivation of X^n in suitable cells results in wild-type phenotype of the fur, and inactivation of X^t results in white patches. It was found that the skin cells in the wild-type areas all had a condensed X^n-chromosome, while those in the white patches had a condensed X^t-chromosome. Ohno and Cattanach comment that their results "may appear too good to be true," as it is known that the pigmentation status of a patch is determined by migrating melanophores and not by the skin cells on which their cytological evidence was based. The two investigators feel, however, that the fact of complete correlation

between the patterns of mosaic coat color and mosaic chromosome condensation in their cases can be explained on the basis
of a close cell-lineage relation of skin and pigment cells.

Limitations of the Single-active-X Hypothesis. In the few years
since the single-active-X hypothesis was proposed, it has proven
itself to be an ingenious interpretation of already known, but
not understood phenomena and a stimulus to asking new questions and obtaining new answers. What more can be expected
of a hypothesis? Moreover, many of the new answers affirm
the validity of the hypothesis. It is no detraction from its value
to add that some of the answers have implied limits to the
hypothesis (L. B. Russell 1964b). Not all X-linked genes in man
and mouse and not all autosome-to-X-chromosome translocations in the latter occur in two alternative cell populations in
heterozygous females. Thus, in women of the heterozygous
X-linked genotype Xg^a/Xg where Xg^a stands for an allele which
causes presence of a specific antigen on the red blood cells and
Xg causes absence of the antigen, only Xg (a +) and no Xg (a −)
red blood cells are found (Gorman et al. 1963; Reed, Simpson,
and Chown 1963). If the Xg^a antigen is produced by the developing red blood cells themselves, as seems likely, then its
presence on all cells signifies that no inactivation of the Xg^a
allele takes place in any of them. This indeed appears to be
the most plausible interpretation, although it is conceivable that
the antigen is produced by other cells (possibly existing as two
alternative populations) and that it coats all red blood cells. In
the mouse, a similar lack of mosaicism in females heterozygous
for an X-linked gene is shown by the mutant scurfy.

Concerning translocation heterozygotes, L. B. Russell (1963)
has shown that the inactivation of autosomal loci translocated
to an X-chromosome depends on their distance from a certain
specific part or parts of the X-chromosome. There is a spreading
effect of inactivation from the "inactivation center," which in
some translocations always, in others only occasionally,
and in still others never, reaches the translocated autosomal
gene. On this basis it would not be unexpected that in normal
X-chromosomes some loci do not get inactivated. The issue is
by no means settled, and Lyon (1966) has presented arguments
for "Lack of evidence that inactivation of the mouse X-chro-

mosome is incomplete." In man, though, it is granted that it may be incomplete indeed, because inactivation of a whole X-chromosome does not readily fit two well-known facts. XO women are morphologically different from XX women, and XXY men different from XY men. If a whole X-chromosome were inactivated in XX and XXY individuals, why would XX not appear like XO and XXY not like XY? Accordingly, at least in man, the single-active-X-chromosome hypothesis may have to be revised to a "single-active-X-chromosome-region" hypothesis. It is in conformity with these genetic indications that, cytologically, only a part of the X-chromosome is heteropycnotic or delayed in replication, but that there are regions which are not heteropycnotic and not delayed in replication and thus behave like most other chromosomes.

If indeed certain X-linked loci are not included in the inactivation process that serves as a dosage-compensation mechanism, they may then be subject to other types of compensation. It has been shown that hemizygous Xg^a males equal homozygous Xg^aXg^a females in strength of antigen (Mann et al. 1962). The mechanism responsible for this gene-dose independence of the product is unknown.

The irregular functional mosaicism of the human female may be relevant to the well-known fact that at all ages the mortality of women is lower than that of men. A genetic interpretation of this fact had long made use of the one-X versus the two-X genotype of the two sexes. It was argued that life-shortening alleles in the hemizygous male would produce their effect without inhibition by more favorable alleles in another X-chromosome. In females, on the contrary, many of the unfavorable alleles in each X-chromosome would be shielded by favorable alleles in the other. This view may still have validity for the genes lying in a section or sections that never are inactivated. In addition, genes in sections in which only one out of two alleles are active in a given cell will appear in two alternative cell populations. Given this situation, the presence of two X-chromosomes in females no longer constitutes a method of shielding unfavorable alleles in each X-chromosome by favorable ones in the other. Yet the single-active-X-chromosome mechanism which bares, in different cells, both alleles of X-chromosomal loci may therewith confer a bonus on women by endowing them with a greater versatility of genetic functioning than men.

Epilogue. The study of mosaics has gone a long way since the miraculous moth hermaphrodite was described two hundred years ago. With each step in the discovery of fundamental properties of cells and of developing organisms new understanding was gained for an interpretation of mosaicism, and, reciprocally, the analysis of mosaics contributed to our understanding of fundamental biological phenomena. At one time, a patch of yellow bristles on an insect which had mostly black bristles could not be subject to fruitful interpretation. Biology was not yet ready for it. In the course of history the patch of yellow bristles provided insight into chromosome behavior and gene action. At one time, a human hermaphrodite could only inspire awe or abhorrence, depending on the mind of the observer. It took the widest range of biological understanding to make such a person an object of rational inquiry.

Much of the work which led to our present knowledge of mosaicism was done during the period of classical genetics. Yet the discoveries that irregular functional mosaicism is a property of all human and of other mammalian females and that, very likely, antibody-forming tissues are multiple mosaics are achievements of most recent years. True, they are concerned with specific groups of organisms and in this sense are more restricted in extent than many areas of molecular biology. They will have to be complemented by molecular interpretations. The variety of organisms, however, and the understanding of even single individuals permits insights which transcend in scope those furnished by the analysis of the threads out of which the tapestry of Life is woven.

REFERENCES

Altenburg, E., and L. S. Browning
  1961 The relatively high frequency of whole-body mutations compared with fractionals induced by X-rays in Drosophila sperm. *Genetics* 46: 203–212.
Austin, C. R.
  1960 Anomalies of fertilization leading to triploidy. *J. Cellular Comp. Physiol.* 56 (*suppl. 1*): 1–15.

Baikie, A. G., W. M. Court Brown, K. E. Buckton, D. G. Harnden, P. A. Jacobs, and I. M. Tough
1960 A possible specific chromosome abnormality in human chronic myeloid leukaemia. *Nature 188:* 1165–1166.

Barr, M. L., and E. G. Bertram
1949 A morphological distinction between neurones of the male and female, and the behaviour of the nucleolar satellite during accelerated nucleoprotein synthesis. *Nature 163:* 676–677.

Barr, M. L., L. F. Bertram, and H. A. Lindsay
1950 The morphology of the nerve cell nucleus, according to sex. *Anat. Record 107:* 283–292.

Barski, G., S. Sorieul, and F. Cornefert
1960 Production dans des cultures in vitro de deux souches cellulaires en association de cellules de caractère "hybride." *Compt. Rend. 251:* 1825–1827.

Beach, F.
1963 Reproductive anatomy and mating behavior of an intersex rat. *J. Mammol. 44:* 57–61.

Beermann, W.
1956 Nuclear differentiation and functional morphology of chromosomes. *Cold Spring Harbor Symp. Quant. Biol. 21:* 217–232.

Benirschke, K., and L. E. Brownhill
1962 Further observations on marrow chimerism in marmosets. *Cytogenetics 1:* 245–257.
1963 Heterosexual cells in testes of chimeric marmoset monkeys. *Cytogenetics 2:* 331–341.

Benirschke, K., and M. M. Sullivan
1966 Corpora lutea in proven mules. *Fertility Sterility 17:* 24–33.

Beutler, E., M. Yeh, and V. F. Fairbanks
1962 The normal human female as a mosaic of X-chromosome activity: Studies using the gene for G-6-PD-deficiency as a marker. *Proc. Nat. Acad. Sci. U.S. 48:* 9–16.

Bhat, N. R.
1949 A dominant mutant mosaic house mouse. *Heredity 3:* 243–248.

Billingham, R. E., L. Brent, and P. B. Medawar
1956 Quantitative studies on tissue transplantation immunity. III. Actively acquired tolerance. *Phil. Trans. Roy. Soc. London, Ser. B 239:* 357–414.

Böök, J. A.
1964 Some mechanisms of chromosome variations and their relation to human malformations. *Eugenics Rev. 56:* 151–157.

Boveri, T.
1888 Über partielle Befruchtung. *Sitzber. Ges. Morphol. Physiol. München* 4: 64–72.
1889 Die Vorgänge der Zellteilung und Befruchtung in ihrer Beziehung zur Vererbungsfrage. *Beitr. Anthropologie Urgeschichte Bayerns* 8: 27–39.
1915 Über die Entstehung der Eugsterschen Zwitterbienen. *Arch. Entwicklungsmech. Organ.* 41: 264–311.

Bridges, C. B.
1921 Triploid intersexes in Drosophila melanogaster. *Science* 54: 252–254.
1925 Haploidy in Drosophila melanogaster. *Proc. Nat. Acad. Sci. U.S.* 11: 706–710.
1930 Haploid Drosophila and the theory of genic balance. *Science* 72: 405–406.
1932 The genetics of sex in Drosophila. In E. Allen, ed., *Sex and Internal Secretions* (Baltimore; Williams & Wilkins), pp. 55–93.
1939 Cytological and genetic basis of sex. In E. Allen, ed., *Sex and Internal Secretions,* 2 ed. (Baltimore: Williams & Wilkins), pp. 15–63.

Brown, S. W.
1966 Heterochromatin. *Science* 151: 417–425.

Cappe de Baillon, P.
1927 Recherches sur la tératologie des insectes. *Encyclopédie Entomologique, Sér. A.* 8: 1–291.

Cattanach, B. M.
1961 A chemically-induced variegated-type position effect in the mouse. *Z. Vererbungslehre* 92: 165–182.

Chu, E. H. Y., H. C. Thuline, and D. E. Norby
1964 Triploid-diploid chimerism in a male tortoiseshell cat. *Cytogenetics* 3: 1–18.

Clarac, F. de
1836–1837 *Musée de Sculpture Antique et Moderne. Planches. Statues,* Vol. 4, plates 545–780. Paris: V. Texier.

Cock, A. G.
1964 Dosage compensation and sex-chromatin in non-mammals. *Genet. Res.* 5: 354–365.

Cooper, K. W.
1959 A bilaterally gynandromorphic Hypodynerus, and a summary of cytologic origins of such mosaic hymenoptera. Biology of eumenine wasps, VI. *Bull. Florida State Museum Biol. Sci.* 5: 25–40.

Court Brown, W. M.
1965 In *Aging and Levels of Biological Organization*, ed. A. M. Brues and G. A. Sacher. Section II. Genetics and Environment, pt. 2, p. 82. Chicago: Univ. of Chicago Press.

Court Brown, W. M., K. E. Buckton, P. A. Jacobs, I. M. Tough, E. V. Kuenssberg, and J. D. E. Knox
1966 Chromosome studies on adults. *Eugenics Lab. Mem. No. 42.* Cambridge, England: Univ. Press, 91 pp.

Crew, F. A. E., and R. Lamy
1938 Mosaicism in *Drosophila pseudoobscura*. *J. Genet. 37:* 211–228.

Crew, F. A. E., and S. S. Munro
1938 Gynandromorphism and lateral asymmetry in birds. *Proc. Roy. Soc. Edinburgh B 58(2):* 114–134.

Danforth, C. H.
1937a An experimental study of plumage in Reeves pheasants. *J. Exptl. Zool. 77:* 1–11.
1937b Artificial gynandromorphism and plumage in *Phasianus*. *J. Genet. 34:* 497–506.

Danforth, J. C., and J. B. Price
1935 Failure of theelin and thyroxin to affect plumage and eye-colour of the blackbird. *Proc. Soc. Exptl. Biol. Med. 32:* 675–678.

Davidson, R. G., H. M. Nitowsky, and B. Childs
1963 Demonstration of two populations of cells in the human female heterozygous for glucose-6-phosphate dehydrogenase variants. *Proc. Nat. Acad. Sci. U.S. 50:* 481–485.

Davidson, W. M., J. F. Fowler, and D. Robertson Smith
1958 Sexing the neutrophil leucocytes in natural and artificial blood chimaeras. *Brit. J. Haematol. 4:* 231–238.

Demerec, M., and H. Slyzinska
1937 Mottled white 258-18 of *Drosophila melanogaster*. *Genetics 22:* 641–649.

Dobzhansky, T.
1930 Studies on the intersexes and supersexes in *Drosophila melanogaster*. (Russian, with English summary.) *Bull. Bur. Genet. 8:* 91–158.
1957 The X-chromosome in the larval salivary glands of hybrids *Drosophila insularis* x *Drosophila tropicalis*. *Chromosoma 8:* 691–698.

Dobzhansky, T., and C. B. Bridges
1928 The reproductive system of triploid intersexes in *Drosophila melanogaster*. *Am. Naturalist 62:* 425–434.

Drescher, W., and W. C. Rothenbuhler
1963   Gynandromorph production by egg chilling. *J. Heredity 54:* 195–201.

Dreyfus, J. C., M. Bessis, and J. C. Kaplan
1964   L'existence de deux populations de globules rouges chez les femmes hétérozygotes pour le déficit en glucose-6-phosphate déshydrogénase. Démonstration par spectrophotométrie sur globules rouges examinés individuellement. *Compt. Rend. 258:*3791–3793.

Dudich, E.
1923   Über einen somatischen Zwitter des Hirschkäfers. *Entomol. Blätter 19:* 129–133.

Dunn, H. O., W. C. Wagner, K. McEntee, W. H. Stone, and J. Caulton
1966   Blood cell chimerism in mature dizygotic twin bulls (*Bos taurus*). *Mammalian Chromosomes Newsletter 19:* 19.

Dunsford, I., C. C. Bowley, A. M. Hutchison, J. S. Thompson, R. Sanger, and R. R. Race
1953   A human blood-group chimera. *Brit. Med. J. 2:* 81.

Emery, A. E. H.
1964   Lyonisation of the X chromosome. *Lancet 1:* 884.

Evans, H. J., C. E. Ford, M. F. Lyon, and J. Gray
1965   DNA replication and genetic expression in female mice with morphologically distinguishable X-chromosomes. *Nature 206:* 900–903.

Fechheimer, N. S., M. S. Herschler, and L. O. Gilmore
1963   Sex chromosome mosaicism in unlike sexed cattle twins. *Proc. Intern. Congr. Genet. 11th (The Hague 1963), Genetics To-day 1:* 265.

Ford, C. E., P. E. Polani, J. H. Briggs, and P. M. F. Bishop
1959   A presumptive human XXY/XX mosaic. *Nature 183:* 1030–1032.

Gartler, S. M., and R. S. Sparkes
1963   The Lyon-Beutler hypothesis and isochromosome X patients with the Turner syndrome. *Lancet 2:* 411.

Gartler, S. M., S. H. Waxman, and E. Giblett
1962   An XX/XY human hermaphrodite resulting from double fertilization. *Proc. Nat. Acad. Sci. U.S. 48:* 332–335.

German, J.
1964   The pattern of DNA synthesis in the chromosomes of human blood cells. *J. Cell Biol. 20:* 37–55.

Giannelli, F.
1963   The pattern of X-chromosome deoxyribonucleic acid syn-

thesis in two women with abnormal sex-chromosome complements. *Lancet 1:* 863–865.

Gilchrist, G. S., D. Hammond, and J. Melnyk
1965 Hemophilia A in a phenotypically normal female with XX/XO mosaicism. *New Engl. J. Med. 273:* 1402–1406.

Gläss, E.
1957 Das Problem der Genomsonderung in den Mitosen unbehandelter Rattenlebern. *Chromosoma 8:* 468–492.
1961 Weitere Untersuchungen zur Genomsonderung I. Modellversuche zum Phänomen der Genomsonderung. *Chromosoma 12:* 410–421.

Goldschmidt, R.
1915 Vorläufige Mitteilung über weitere Versuche zur Vererbung und Bestimmung des Geschlechts. *Biol. Zentr. 35:* 565–570.
1934 Lymantria. *Bibliog. Genet. 11:* 1–180.

Goldschmidt, R., and K. Katsuki
1928 Cytologie des erblichen Gynandromorphismus von *Bombyx mori* L. *Biol. Zentr. 48:* 685–699.
1931 Vierte Mitteilung über erblichen Gynandromorphismus und somatische Mosaikbildung bei *Bombyx mori* L. *Biol. Zentr. 51:* 58–74.

Gorman, J. G., J. Dire, A. M. Treacy, and A. Cahan
1963 The application of −Xg^a antiserum to the question of red cell mosaicism in female heterozygotes. *J. Lab. Clin. Med. 61:* 642–649.

Grüneberg, H.
1966 The case for somatic crossing-over in the mouse. *Genet. Res. 7:* 58–75.
1967 Sex-linked genes in man and the Lyon hypothesis. *Ann. Human Genet. 30:* 239–257.

Hannah-Alava, A.
1960 Genetic Mosaics. *Sci. Am.* (May): 119–130.

Hayman, D. L., and P. G. Martin
1965 Sex chromosome mosaicism in the marsupial genera Isoodon and Perameles. *Genetics 52:* 1201–1206.

Hildreth, P. E.
1965 Doublesex, a recessive gene that transforms both males and females of Drosophila into intersexes. *Genetics 51:* 659–678.

Hollander, W. F.
1949 Bipaternity in pigeons. *J. Heredity 40:* 271–277.

Hollander, W. F., J. W. Gowen, and J. Stadler
1956 A study of 25 gynandromorphic mice of the Bagg albino strain. *Anat. Record 124:* 223-242.

Hughes-Schrader, S.
1927 Origin and differentiation of the male and female germ cells in the hermaphrodite of *Icerya purchasi* (Coccidae). *Z. Zellforsch. Mikroskop. Anat. 6:* 509-540.

Hughes-Schrader S., and D. F. Monahan
1966 Hermaphroditism in *Icerya zeteki* Cockerell, and the mechanism of gonial reduction in iceryine coccids (Coccoidea: Margarodidae Morrison). *Chromosoma 20:* 15-31.

James, A. P.
1955 A genetic analysis of sectoring in ultraviolet-induced variant colonies of yeast. *Genetics 40:* 204-213.

Kadowaki, J.-I., W. W. Zuelzer, A. J. Brough, R. I. Thompson, P. V. Woolley, Jr. and D. Gruber
1965 XX/XY lymphoid chimaerism in congenital immunological deficiency syndrome with thymic alymphoplasia. *Lancet 2:* 1152-1156.

Käfer, E.
1961 The processes of spontaneous recombination in vegetative nuclei of *Aspergillus nidulans. Genetics 46:* 1581-1609.

Keck, W. N.
1934 The control of the secondary sex characters in the English sparrow, *Passer domesticus* (Linnaeus). *J. Exptl. Zool. 67:* 315-347.

Kerkis, J. J.
1934 On the mechanism of the development of triploid intersexuality in *Drosophila melanogaster. Compt. Rend. Acad. Sci. U.S.S.R.* n.s. 3: 288-294.

Klarwill, V. von, ed.
1924 *The Fugger News-Letters: Being a Selection of Unpublished Letters from the Correspondents of the House of Fugger during the Years 1568-1605.* London: John Lane the Bodley Head Ltd. 284 pp.

Kline, A. H., J. B. Sidbury, Jr., and C. P. Richter
1959 The occurrence of ectodermal dysplasia and corneal dysplasia in one family. *J. Pediat. 55:* 355-366.

Klinger, H. P., and H. G. Schwarzacher
1962 XY/XXY and sex chromatin positive cell distribution in a 60 mm human fetus. *Cytogenetics 1:* 266-290.

Knudson, A. G., Jr.
1965   Genetics and Disease. New York: McGraw-Hill. 294 pp.

Laidlaw, H. H., and K. W. Tucker
1964   Diploid tissue derived from accessory sperm in the honey bee. Genetics 50: 1439-1442.

Lederberg, J.
1959   Genes and antibodies. Science 129: 1649-1653.

Lejeune, J.
1964   The 21 trisomy—current stage of chromosomal research. Progr. Med. Genet. 3: 144-177.

Lejeune, J., J. Lafourcade, K. Schärer, E. de Wolff, C. Salmon, M. Haines, and R. Turpin
1962   Monozygotisme hétérocaryote, jumeau normal et jumeau trisomique 21. Compt. Rend. 254: 4404-4406.

Lejeune, J., and R. Turpin
1961   Détection chromosomique d'une mosaique artificielle humaine. Compt. Rend. 252: 3148-3150.

Lengerken, H. v.
1928   Über die Entstehung bilateral-symmetrischer Insektengynander aus verschmolzenen Eiern. Biol. Zentr. 48: 475-509.

Lillie, F. R.
1917   The free-martin; a study of the action of sex hormones in the foetal life of cattle. J. Exptl. Zool. 23: 371-452.

Linder, D., and S. M. Gartler
1965a Distribution of glucose-6-phosphate dehydrogenase electrophoretic variants in different tissues of heterozygotes. Am. J. Human Genet. 17: 212-220.
1965b Glucose-6-phosphate dehydrogenase mosaicism: Utilization as a cell marker in the study of leiomyomas. Science 150: 67-69.

Lindsten, J.
1963   The Nature and Origin of X Chromosome Aberrations in Turner's Syndrome. Stockholm: Almqvist & Wiksell. 167 pp.

London, D. R., N. H. Kemp, J. R. Ellis, and U. Mittwoch
1964   Turner's syndrome with secondary amenorrhoea and sex chromosome mosaicism. Acta Endocrinol. 46: 341-351.

Lukas, J. G.
1803-1804   Vermischte Beyträge zur Fortschreitung der Wissenschaft in der Bienenzucht. Leipzig: Fr. Fleischer.

Lyon, M. F.
1961   Gene action in the X-chromosome of the mouse (Mus musculus L.). Nature 190: 372-373.
1962   Sex chromatin and gene action in the mammalian X-chromosome. Am. J. Human Genet. 14: 135-148.

1966 Lack of evidence that inactivation of the mouse X-chromosome is incomplete. *Genet. Res. 8:* 197–203.

Lyon, M. F., A. G. Searle, C. E. Ford, and S. Ohno
1964 A mouse translocation suppressing sex-linked variegation. *Cytogenetics 3:* 306–323.

MacKnight, R. H.
1937 Haploid ovarian tissue in Drosophila. *Proc. Nat. Acad. Sci. U.S. 23:* 440–443.

McKusick, V. A.
1962 On the X chromosome of man. *Quart. Rev. Biol. 37:* 69–175.

Makino, S., J. Muramoto, and T. Ishikawa
1965 Notes on XX/XY mosaicism in cells of various tissues of heterosexual twins of cattle. *Proc. Japan Acad. 41:* 414–418.

Mangold, O., and F. Seidel
1927 Homoplastische und heteroplastische Verschmelzung ganzer Tritonkeime. *Arch. Entwicklungsmech. Organ. 111:* 593–665.

Mann, J. D., A. Cahan, A. G. Gelb, N. Fisher, J. Hamper, P. Tippett, R. Sanger, and R. R. Race
1962 A sex-linked blood group. *Lancet 1:* 8–10.

Melander, Y.
1950 Chromosome behaviour of a triploid adult rabbit as produced by Häggqvist and Bane after colchicine treatment. *Hereditas 36:* 335–341.

Mikkelsen, M., A. Frøland, and J. Ellebjerg
1963 XO/XX mosaicism in a pair of presumably monozygotic twins with different phenotypes. *Cytogenetics 2:* 86–98.

Mintz, B.
1965 Genetic mosaicism in adult mice of quadriparental lineage. *Science 148:* 1232–1233.

Mohr, O. L.
1932 Genetical and cytological proof of somatic elimination of the fourth chromosome in Drosophila melanogaster. *Genetics 17:* 60–80.

Money, J.
1963 Cytogenetic and psychosexual incongruities with a note on space-form blindness. *Am. J. Psychiat. 119:* 820–827.

Morgan, L. V.
1929 Composites of Drosophila melanogaster. *Carnegie Inst. Wash. Publ. 399:* 223–296.

Morgan, T. H.
1905 An alternative interpretation of gynandromorphous insects. *Science 21:* 1–6.

Morgan, T. H., and C. B. Bridges
1919 Contributions to the genetics of Drosophila melanogaster. I. The origin of gynandromorphs. Carnegie Inst. Wash. Publ. 278: 1–122.

Morris, J. M., and V. B. Mahesh
1963 Further observations on the syndrome, "testicular feminization." Am. J. Obstet. Gynecol. 87: 731–748.

Mukherjee, A. S.
1966 Dosage compensation in Drosophila: An autoradiographic study. Nucleus (Calcutta) 9: 83–96.

Mukherjee, A. S., and W. Beermann
1965 Synthesis of ribonucleic acid by the X-chromosome of Drosophila melanogaster and the problem of dosage compensation. Nature 207: 785–786.

Mukherjee, B. B., and A. K. Sinha
1964 Single-active-X hypothesis: Cytological evidence for random inactivation of X-chromosomes in a female mule complement. Proc. Nat. Acad. Sci. U.S. 51: 252–259.

Muldal, S., C. W. Gilbert, L. G. Lajthe, J. Lindsten, J. Rowley, and M. Fraccaro
1963 Tritiated thymidine incorporation in an isochromosome for the long arm of the X chromosome in man. Lancet 1: 861–863.

Muller, H. J.
1930 Types of visible variations induced by X-rays in Drosophila. J. Genet. 22: 299–334.
1932 Further studies on the nature and causes of gene mutations. Proc. Intern. Congr. Genet. 6th (Ithaca, N.Y. 1932) 1: 213–255.
1950 Evidence of the precision of genetic adaptation. Harvey Lectures, Ser. 43 (1947–1948): 165–229.

Muller, H. J., and W. D. Kaplan
1966 The dosage compensation of Drosophila and mammals as showing the accuracy of the normal type. Genet. Res. 8: 41–60.

Muller, H. J., B. B. League, and C. A. Offermann
1931 Effect of dosage changes of sex-linked genes, and the compensatory effect of other gene differences between male and female. (Abstr.) Anat. Record 51 (suppl.): 110.

Nowell, P. C., and D. A. Hungerford
1961 Chromosome studies in human leukemia. II: Chronic granulocytic leukemia. J. Nat. Cancer Inst. 27: 1013–1035.

Offermann, C. A.
1936 Branched chromosomes as symmetrical duplications. *J. Genet.* 32: 103–116.

Ohno, S., and B. M. Cattenach
1962 Cytological study of an X-autosome translocation in *Mus musculus*. *Cytogenetics* 1: 129–140.

Ohno, S., and A. Gropp
1965 Embryological basis for germ cell chimerism in mammals. *Cytogenetics* 4: 251–261.

Ohno, S., J. Jainchill, and C. Stenius
1963 The creeping vole (*Microtus oregoni*) as a gonosomic mosaic. I: The OY/XY constitution of the male. *Cytogenetics* 2: 232–239.

Ohno, S., W. D. Kaplan, and R. Kinosita
1959 Formation of the sex chromatin by a single X-chromosome in liver cells of *Rattus norvegicus*. *Exptl. Cell Res.* 18: 415–418.

Ohno, S., and S. Makino
1961 The single-X nature of sex chromatin in man. *Lancet* 1: 78–79.

Ohno, S., J. Trujillo, C. Stenius, L. C. Christian, and R. Teplitz
1962 Possible germ cell chimeras among newborn dizygotic twin calves (*Bos taurus*). *Cytogenetics* 1: 258–265.

Owen, R. D.
1945 Immunogenetic consequences of vascular anastomoses between bovine twins. *Science* 102: 400–401.

Pearson, C. M., W. M. Fowler, and S. W. Wright
1963 X-chromosome mosaicism in females with muscular dystrophy. *Proc. Nat. Acad. Sci. U.S.* 50: 24–31.

Pearson, N. E.
1927 A study of gynandromorphic katydids. *Am. Naturalist* 61: 283–285.

Pontecorvo, G., and J. A. Roper
1953 Diploid and mitotic recombination. In Pontecorvo, G., The genetics of *Aspergillus nidulans*. *Advan. Genet.* 5: 141–238, (pp. 218–233).

Race, R. R., and R. Sanger
1962 *Blood Groups in Man,* 4 ed. Philadelphia: F. A. Davis Co.

Reed, T. E., and H. F. Falls
1955 A pedigree of aniridia with a discussion of germinal mosaicism in man. *Am. J. Human Genet.* 7: 28–38.

Reed, T. E., N. E. Simpson, and B. Chown
1963 The Lyon hypothesis. *Lancet* 2: 467–468.

Rothenbuhler, W. C.
1958   Genetics and the breeding of the honey bee. *Ann. Rev. Entomol. 3:* 161–180.

Russell, L. B.
1961   Genetics of mammalian sex chromosomes. *Science 133:* 1795–1803.
1963   Mammalian X-chromosome action: Inactivation limited in spread and in region of origin. *Science 140:* 976–978.
1964a Genetic and functional mosaicism in the mouse. In *Role of Chromosomes in Development* (New York: Academic Press) pp. 153–181.
1964b Another look at the single-active-X hypothesis. *Trans. N.Y. Acad. Sci. 26:* 726–736.

Russell, L. B., and J. W. Bangham
1959   Variegated-type position effects in the mouse (Abstr.). *Genetics 44:* 532.
1961   Variegated-type position effects in the mouse. *Genetics 46:* 509–525.

Schäffer, J. C.
1761   *Der wunderbare und vielleicht in der Natur noch nie erschienene Eulenzwitter.* Regensburg.

Schrader, F., and A. H. Sturtevant
1923   A note on the theory of sex determination. *Am. Naturalist 57:* 379–381.

Seiler, J.
1949   Das Intersexualitätsphänomen. *Experientia 5:* 425–438.
1958   Die Entwicklung des Genitalapparates bei triploiden Intersexen von *Solenobia triquetrella* F.R. (*Lepid. Psychidae*). Deutung des Intersexualitätsphänomens. *Arch. Entwicklungsmech. Organ. 150:* 199–372.
1965   Sexuality as developmental process. *Proc. Intern. Congr. Genet. 11th (The Hague 1963) Genetics To-day 2:* 199–207.

Siebold, C. T. v.
1864   Ueber Zwitterbienen. *Z. wiss. Zool. 14:* 73–80.

Smithies, O.
1963   Gamma-globulin variability: a genetic hypothesis. *Nature 199:* 1231–1236.

Sorieul, S., and B. Ephrussi
1961   Karyological demonstration of hybridization of mammalian cells in vitro. *Nature 190:* 653–654.

Stern, C.
1929   Über die additive Wirkung multipler Allele. *Biol. Zentr. 49:* 261–290.

1936   Somatic crossing over and segregation in Drosophila melano-
        gaster. Genetics 21: 625–730.
1960a   Dosage compensation—development of a concept and new
        facts. Can. J. Genet. Cytol. 2: 105–118. Reprinted in M. F.
        Ashley Montagu, ed., Genetic Mechanisms in Human Disease:
        Chromosomal Aberrations (Springfield: C. C. Thomas, 1961),
        pp. 524–544.
1960b   A mosaic of Drosophila consisting of 1X, 2X and 3X tissue
        and its probable origin by mitotic non-disjunction. Nature
        186: 179–180.
1960c   Principles of Human Genetics, 2 ed. San Francisco: W. H.
        Freeman Co. Copyright © 1960. 753 pp.
1966   Pigmentation mosaicism in intersexes of Drosophila. Rev.
        Suisse Zool. 73: 339–355.

Stern, C., and K. Sekiguti
1931   Analyse eines Mosaikindividuums bei Drosophila melano-
        gaster. Biol. Zentr. 51: 194–199.

Stone, W. H., D. T. Berman, W. J. Tyler, and M. R. Irwin
1960   Blood types of the progeny of a pair of cattle twins showing
        erythrocyte mosaicism. J. Heredity 51: 136–140.

Stone, W. H., J. Friedman, and A. Fregin
1964   Possible somatic cell mating in twin cattle with erythrocyte
        mosaicism. Proc. Nat. Acad. Sci. U.S. 51: 1036–1044.

Stormont, C., W. C. Weir, and L. L. Lane
1953   Erythrocyte mosaicism in a pair of sheep twins. Science 118:
        695–696.

Sturtevant, A. H.
1920   Intersexes in Drosophila simulans. Science 51: 325–327.
1929   The claret mutant type of Drosophila simulans: a study of
        chromosome elimination and of cell-lineage. Z. wiss. Zool.
        135: 323–356.

Tandler, J., and K. Keller
1911   Über das Verhalten des Chorions bei verschiedengeschlecht-
        licher Zwillingsgravidität des Rindes und über die Morpho-
        logie des Genitales der weiblichen Tiere, welche einer solchen
        Gravidität entstammen. Deut. tierärztl. Wóchschr. 19: 148–149.

Tarkowski, A. K.
1961   Mouse chimaeras developed from fused eggs. Nature 190:
        857–860.
1963   Studies on mouse chimeras developed from eggs fused in
        vitro. Nat. Cancer Inst. Monograph 11: 51–71.
1964a   Patterns of pigmentation in experimentally produced mouse
        chimaerae. J. Embryol. Exptl. Morphol. 12: 575–585.

1964b True hermaphroditism in chimaeric mice. *J. Embryol. Exptl. Morphol.* 12: 735-757.

Taylor, J. H.
1960 Asynchronous duplication of chromosomes in cultured cells of Chinese hamster. *J. Biophys. Biochem. Cytol.* 7: 455-464.

Tjio, J. H., and A. Levan
1956 The chromosome number of man. *Hereditas* 42: 1-6.

Turpin, R., J. Lejeune, J. Lafourcade, P. L. Chigot, and C. Salmon
1961 Présomption de monozygotisme en dépit d'un dimorphisme sexuel: sujet masculin XY et sujet neutre Haplo X. *Compt. Rend.* 252: 2945-2946.

Uchida, I. A., H. C. Wang, and M. Ray
1964 Dizygotic twins with XX/XY chimerism. *Nature* 204: 191.

Vandenberg, S. G., V. A. McKusick, and A. B. McKusick
1962 Twin data in support of the Lyon hypothesis. *Nature* 194: 505-506.

Wald, H.
1936 Cytologic studies on the abnormal development of the eggs of the claret mutant type of *Drosophila simulans*. *Genetics* 21: 264-281.

Welshons, W. J., and L. B. Russell
1959 The Y-chromosome as the bearer of male determining factors in the mouse. *Proc. Nat. Acad. Sci. U.S.* 45: 560-566.

Wheeler, W. M.
1910 A gynandromorphous Mutillid. *Psyche* 17: 186-190.

Whiting, P. W.
1932 Reproductive reactions of sex mosaics of a parasitic wasp, *Habrobracon juglandis*. *J. Comp. Psych.* 14: 345-363.
1940 Multiple alleles in sex determination of Habrobracon. *J. Morphol.* 66: 323-355.

Whiting, P. W., H. J. Greb, and B. R. Speicher
1934 A new type of sex-intergrade. *Biol. Bull.* 66: 153-165.

Witschi, E.
1948 Migration of the germ cells of human embryos from the yolk sac to the primitive gonadal folds. *Carnegie Inst. Wash. Publ. 575, Contributions to Embryology* 32: 76-80.

Woodiel, F. N., and L. B. Russell
1963 A mosaic formed from fertilization of two meiotic products of oogenesis. (Abstr.) *Genetics* 48: 917.

Yerganian, G.
1959 Chromosomes of the Chinese hamster, *Cricetulus griseus*. I.

The normal complement and identification of sex chromosomes. *Cytologia (Tokyo)* 24: 66–75.

Zuelzer, W. W., K. M. Beattie, and L. E. Reisman
1964   Generalized unbalanced mosaicism attributable to dispermy and probable fertilization of a polar body. *Am. J. Human Genet.* 16: 38–51.

Developmental Genetics of Pattern

Die Natur, unser Mass,
Hebt sich in Vielheit empor.
Muster lässt sie entsteh'n,
Einheit leuchtet hervor.

—Anonymous

Genetics and Experimental Embryology. During the first quarter of the present century genetics and experimental embryology were among the most successful biological fields. It was obvious that there must be a close interrelation between the genetic determinants of phenotypes and the differentiation of the cells of an embryo, but it was difficult to unite the different approaches of the study of genes and of development into a harmonious discipline. Before 1900, when Mendel's work was rediscovered, August Weismann and, in less extreme form, Wilhelm Roux had speculated on the mode by means of which the totipotent fertilized egg can differentiate into the variety of cell types, tissues, and organs, each of which represent a selected sample of potencies. The hypothesis of these men that the basis of differentiation is an actual sorting out of potencies never was very likely and finally was shown to be invalid. Driesch's proof that the early blastomeres of sea-urchin eggs are still totipotent—that, in his terminology, the prospective significance of these cells is not an indicator of their potencies—was the beginning of a series of demonstrations which showed that differentiation must be understood in terms of differential genic action and not of differential genic content. Most recently, the totipotency of differentiated cells has been shown particularly impressively in experiments with cells derived from roots, hypocotyl, and embryos of wild carrots (review in Torrey, 1966). A hundred thousand cells grown from a single embryo are able to develop separately into embryoids and each embryoid is able to develop into a whole carrot plant (Fig. 39).

It was Driesch, then only 27 years old, who in 1894 prophetically formulated an alternative theory to that which postulated somatic genic segregation as the mechanism of differentiation. In his *Analytische Theorie der Organischen Entwicklung* he wrote as follows: "We assume the nucleus to be a mixture of enzyme-like substances, each of which represents a kind of elementary process of embryonic development . . . Any specific enzyme of this mixture is activated by a specific release mechanism . . . This hypothesis includes that of retention of nuclear totality in spite of nuclear diversity."

Differentiation alone does not account for the orderliness of development. This orderliness depends on the fact that differentiation occurs in a patterned fashion, producing brain here, intestine there, each organ and tissue at its appropriate

FIG. 39. The production of whole wild carrot plants (*Daucus carota*) from free cells derived from the phloem or from embryos. (From Steward 1964. Copyright 1964 by the American Association for the Advancement of Science.)

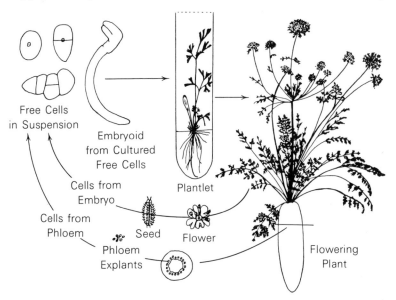

Free Cells
in Suspension

Embryoid
from Cultured
Free Cells

Cells from
Embryo

Cells from
Phloem

Plantlet

Seed

Flower

Phloem
Explants

Flowering
Plant

place. Driesch's hypothesis of differential genic action included concepts which accounted not only for the fact of differentiation but also for its orderly, patterned nature. E. B. Wilson and Oskar Hertwig likewise envisaged schemes for the origin of patterned differentiation. The decisive discovery, however, came only in 1924 when Spemann and Hilde Mangold showed that the localization of the differentiation of amphibian ectoderm into the central nervous system depends on the localization of the organizer which, in the form of the grey crescent, is preexistent in the uncleaved egg as a cytoplasmatic singularity. In a corresponding manner, the localized differentiation of other tissues and organs was traceable to a pattern of regularly distributed embryonic inductors.

At this stage, the genetic basis of development seemed remote if not irrelevant to the approach of the experimental embryologist. When in 1936 Spemann summarized his life's work in the volume *Experimentelle Beiträge zu einer Theorie der Entwicklung,* and when in 1939 Paul Weiss published his penetrating *Principles of Development,* neither of them seemed to require the concept of the gene for their presentation. Indeed, one will look in vain in these books for the word *gene.*

Transplantation Mosaics. Even though the concepts of genetics did not enter deeply into the work of the experimental embryologists in the period just discussed, important use was made of genetically different eggs. Fritz Baltzer and his students, particularly Hadorn, investigated the interaction in development between nucleus and cytoplasm by combining the nucleus of one species of amphibians with the cytoplasm of another. Spemann produced mosaics by combining parts of embryos of the two different newt species, *Triton cristatus* and *T. taeniatus,* whose cells differ in pigmentation. This was useful for following the fate of the vitally "self-stained" parts, but the genetic difference between the species played no further role. The eggs of both species form a primary organizer, and the cells of both respond equally to the inducing stimulus.

A genetically more informative type of experiment consisted in producing transplantation mosaics between urodeles and anurans, such as newts and frogs (review in Spemann, 1936, 1938, chap. 17). The embryos of these two amphibian orders are still alike in their ability to induce differentiation according to a common general pattern, and the cells of one are still able to respond to the inducing stimulus of the other. The results of this response, however, are fundamentally different according to the species used. Normally, in the embryonic mouth region the frog ectoderm differentiates suckers and horny denticles in contrast to the newt ectoderm, which differentiates balancers and dentine teeth. In a transplantation mosaic, frog-belly ectoderm transplanted to the mouth region of a newt embryo is induced to form head structures characteristic of a tadpole, and newt-belly ectoderm transplanted to the mouth region of a frog embryo is induced to form newt-specific structures. Thus, the genotype of the reacting cells is decisive. While the stimulus for differentiation comes from the genetically foreign cells, the type of reaction to it is species-specific, "autonomous." In his lectures in this country Spemann liked to use an analogy for the behavior of his amphibian chimeras. An American private, he imagined, if asked by a German officer to salute him, would say: "Yes, I shall respond to your command, but I shall salute you in my own American way."

Pattern Genes in Drosophila. The experiments with amphibian mosaics are rightly famous, but the genetic differences between the forms whose tissues were combined were not subject to analysis. Progress seemed to depend on the use of simpler genetic differences accounting for diverse patterns of development. Such differences are available in Drosophila. Of course, many of its genes are not pattern determining, but exert their effect wherever the differentiation of the fly permits their expression. Thus the pigmentation genes for yellow, black, and ebony produce their color more or less over the whole body surface, or the genes for singed and forked, which determine the shapes of bristles, do so wherever a bristle is differentiating. Even genes for such traits as type of eye color are not pattern determining. While their effect primarily becomes visible in the eyes only, their localized expression is a consequence of the localized differentiation of eyes.

Besides genes that act without entering into the production of patterns, there are other genes that may be designated as pattern genes. They control the onset of differentiation in specific regions. Among such genes, those that determine the development of the bristles have played an important role in developmental genetics. The bristles are sensory organs composed of four cells, the products of two terminal divisions of a parent cell. One of the four cells forms the bristle proper, another the socket surrounding its base, a third is a peripheral nerve cell, and the fourth provides a nerve sheath. Other genes being normal, a wild-type allele at the so-called scute locus (sc) assures the presence of 11 pairs of bristles on the dorsal mesothorax, each one at a highly constant location, and all or most of specific length and maximal width. Two other pairs of bristles are differentiated at constant locations on the dorsal prothorax and 7 pairs are differentiated on the head (Fig. 40, sc^+). Many mutant alleles of the scute gene are known, thanks particularly to the work of Russian geneticists published over three decades ago (e. g. Dubinin 1929; Serebrovsky 1930). These alleles control the bristle pattern in characteristic ways. Instead of causing or permitting the appearance of bristles at all normal sites, they lead to their regular appearance at some sites only (Fig. 40). Thus, leaving aside minor variability, it may be stated

---

FIG. 40. Dorsal view of head and thorax of *Drosophila melanogaster*. Solid circles, sites of macrochaetae. Open circles, absence of macrochaetae. sc^+, wild type; sc^5, sc^6, sc^2, sc^1, mutant genotypes. The "shoulders" in these diagrams are the prothorax, and the rest of the visible thorax is the mesothorax. The triangular posterior area of the mesothorax is the scutellum. (From Stern 1965.)

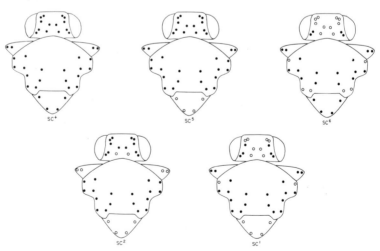

that the allele sc^5 allows the fly to differentiate all but the 2 pairs of scutellar bristles; sc^6 allows the formation of the scutellars but not that of 5 pairs of head bristles and 3 pairs of thoracic bristles; sc^2 again suppresses the formation of the scutellars and in addition that of 3 other pairs on head and prothorax; and sc^1 is responsible for nondifferentiation of the scutellars, 1 other mesothoracic pair, and 4 pairs of head bristles. Similarly, the mutant gene achaete (*ac*), located adjacent to scute, inhibits the appearance of a well-defined assortment of bristles, particularly the dorsocentrals in the middle of the mesothorax.

Several investigators attempted to derive from these facts a developmental interpretation of the origin of bristle patterns. They thought in terms of gradients of bristle-forming potencies of tissues or of streams of determination spreading out from one or several centers. According to these hypotheses, the different mutant genes would control the gradients or they would limit the spread of determination. The facts, however, refused to be forced into these constructs. Whenever a specific model was formulated for the pattern-forming action of one allele, it could not be fitted to the action of another. It had to be concluded that each allele endows the larval imaginal discs which metamorphose into the head and thorax surfaces of the adult with the ability to produce a specific pattern of localized bristle-differentiation.

A given pattern implies a coordinated system, a wholeness of configuration superimposed on the differentiating cell mass. How then can genes influence this wholeness? It seemed reasonable to assume that the type of wholeness is different with each type of allele. This may be illustrated by a hypothetical model. The larval imaginal discs are made up of cell layers that are folded in complex ways. Let us postulate that differentiation of bristles occurs at those points at which folds cut across each other. According to this hypothesis, an allele that leads to differentiation of a specific bristle would be involved in provoking the formation of specific folds. Another allele whose phenotypic effect does not include formation of the bristle would be responsible for a different kind of folding of the imaginal disc. The different types of folding of the discs would constitute different patterns. Since these patterns would precede the appearance of their corresponding bristle patterns, I refer to them as prepatterns. More will be said later about this concept. For the present, the interpretation of the action of the different scute alleles as postulated may be rephrased by saying that different alleles would cause the production of different prepatterns.

Mosaics in Drosophila. Such considerations—independent of the formulation in terms of folding of imaginal discs, which was made solely for the sake of providing a model—led to experiments in which mosaic imaginal discs were produced. If the different alleles indeed cause different prepatterns of the discs, one could expect new types of such prepatterns in mosaic discs, some of whose parts contain one kind of prepattern-forming allele and whose other parts contain an allele responsible for a different prepattern. The new types of overall organization should result in new bristle patterns, and one could hope that the analysis of such new patterns would lead to some understanding of the control of prepattern in genetically homogeneous individuals.

The genetics of pattern is so large a subject that a comprehensive treatment would go far beyond the intention of the present report, whose purpose is limited to telling mainly about the work of our group at Berkeley. We have exploited mosaicism in Drosophila as a means of pattern analysis. There are other types of studies of bristle patterns not based on mosaic analysis; the studies of Turing (1952) and Wigglesworth (1940, 1959) are important examples. Maynard-Smith and Sondhi (1960, 1961), Sondhi (1962a, b, 1963), Hollingsworth (1964), Claxton (1964), and Scharloo, Hoogmoed, and Vreezen (1966) in different approaches have made use of concepts derived from mosaic analysis, and general reviews have been contributed by Ursprung (1963, 1966) and von Borstel (1963).

Mosaic imaginal discs in Drosophila can be produced by combining partially dissociated discs of different mutants or species (Hadorn, Anders, and Ursprung 1959; Nöthiger 1964). For our experiments they were obtained by genetic methods. Let us consider mosaics for the achaete gene. It is located in the X-chromosome. Females which carry a normal allele in one X and a mutant allele in the other are phenotypically normal, the mutant allele being recessive. Gynanders in which the female parts are heterozygous as just described, but in which the male parts carry the mutant allele in their sole X-chromosome, can provide mosaic discs equivalent to transplantation mosaics.

Spontaneously occurring gynanders in Drosophila are usually rare. Strains are available, however, in which a specific X-chromosome is present that is ring-shaped instead of rod-shaped and that is frequently lost from early cleavage nuclei. In an egg that started development with two X-chromosomes, loss of an X in one of two or more cleavage nuclei gives rise to gynandric 2X/1X mosaics. If the rod X carries a recessive pattern gene, for example achaete (ac), and the ring X carries its dominant normal allele (+), then the female parts of the gynander are ac/+ and the male parts ac. In appearance these gynanders vary from one specimen to the next. In one a given imaginal disc may be exclusively XX, in another wholly X, and in still others it may be a mixture of XX and X cells unique for each case in relative size and location. This variability is the product of several independent variables. Both the time and place of chromosome elimination differ from one egg to the other. Furthermore, the descendents of a specific cleavage nucleus or of an embryonic cell form areas that differ from case to case: cell lineage is not fully fixed. This variability enables the investigator to select among the many those relatively few gynanders that are suitable for his analysis.

There is one more point that should be related before we return to the pattern problem. It would not be appropriate to infer the mosaic type of a disc from a discovery of a peculiar bristle pattern. Rather, the mosaic makeup must be ascertained independently of the pattern and, once determined, be used to explain the bristle distribution. The independent determination is made possible by the use of marker genes. If, for instance, the X-chromosome carrying the mutant achaete allele also is made to carry the equally recessive gene for yellow body and bristle coloration (y), and if the other X-chromosome carries the normal allele for bristle pattern and that for dark, not-yellow coloration, then the two parts of gynanders that differ in their bristle genotype are marked by different coloration: dark in 2X and yellow in 1X parts.

Autonomy of Bristle Production. When the bristle patterns of the mosaic individuals were observed, no new patterns were encountered. Evidently no new types of overall organization resulted from the interactions of supposedly different prepatterns in genetically mosaic discs. Rather, it became clear that the decision whether a bristle organ is or is not differentiated at a specific location is made at the location itself, depending on the genotype of the tissue present there (Stern 1954*a, b,* Roberts 1961 for achaete; Stern and Swanson 1957, Young and Lewontin 1966 for scute). If the normal genotype of an unmixed disc would lead to bristle differentiation at a given location, the same genotype would result in differentiation of the bristle even in a mosaic disc as long as the tissue in this location carried the relevant genotype, and regardless of how large the area of the mosaic without the normal genotype. Conversely, if the mutant genotype in an unmixed disc failed to result in bristle differentiation at a given location, it would equally fail to do so as long as the tissue in the location carried the mutant genotype and regardless of how large the area of the mosaic without the mutant genotype (Figs. 41, 42).

To use a simple term, one may state that the different areas

FIG. 41. Mosaic half-thoraxes of *Drosophila melanogaster.* The black areas are hemizygous for the recessive achaetae (*ac*) which causes absence of the posterior dorsocentral bristle (pdc) whose site is indicated above by a solid circle and below by an empty circular space. The white areas are heterozygous for the dominant normal *ac*+ and the *ac* allele. *Above,* four mosaics in which the site of the pdc lies within an *ac/ +* area; *below,* four mosaics in which the site lies within an *ac* area. The symbol + indicates presence, − absence of pdc differentiation. (From Stern 1954*b.*)

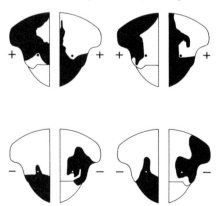

of mosaic discs determine their differentiation autonomously. As a matter of fact, autonomous differentiation of bristles in flies mosaic for scute alleles had been established long ago by Sturtevant (1932), but it had not been properly appreciated. In his pioneering report before the Sixth International Congress of Genetics on "The Use of Mosaics in the Study of the Relation between Genes and Development," Sturtevant presented so many new findings on both autonomy and nonautonomy of gene expression that the significance of autonomy in pattern formation did not impress itself on the reader. Autonomy in pattern, however, seemed to me a surprising paradox. Autonomy of gene expression in mosaics when it relates to nonpatterned phenotypes is easily understood. If, for instance, a mosaic is yellow in genetically yellow parts and not-yellow in genetically not-yellow parts, this autonomous pigmentation simply means that the cells of the two areas are able to decide individually whether to produce one or the other color. Or, if mosaics could be obtained for the normal and a mutant genotype, such as apparently present in one of Hadorn's (1965) cell

---

FIG. 42. Dorsal views of head of *Drosophila melanogaster*. (A) Normal head with micro- and macrochaetae. (B), (C), (D) Heads of three gynandric mosaics (eyes omitted). Solid circles, ocellar bristles; broken circles, absence of ocellar bristles; open circles, bristles other than ocellars in $sc^1/+$ tissue; crossed circles, bristles other than ocellars in sc^1 tissue. (B), (C) background female, $sc^1/+$; male patch, sc^1. Autonomous nonformation of ocellar bristles in the sc^1 area. (D) background male, sc^1, female patch $sc^1/+$. Autonomous formation of ocellar bristle in the $sc^1/+$ area. (From Stern and Swanson 1957.) The pattern of bristle differentiation in this experiment with sc^1 differs somewhat from that diagramatically shown in Fig. 40.

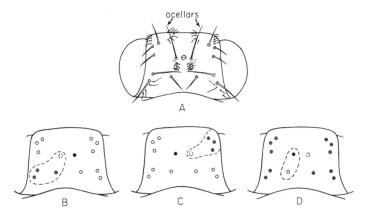

cultures, which was unable to form any bristles anywhere, then the mosaics would very likely result in a phenotypic combination of bristled and bristleless areas. On the other hand, by definition, the production or nonproduction of a bristle in a pattern mutant seemed to require a decision not solely made on the spot, but made as a result of prepattern organizations in which the fate of a specific cell depends on its position within the whole.

I remember well how I brooded over my mosaics that showed the paradoxical autonomy in patterning until I suddenly realized that the paradox was not in Nature but in my preconceived views. Instead of pretending "The facts do not fit my theory—the worse for the facts," or complaining, "Life has double-crossed the . . . writer," I had to admit the wisdom of Goethe: "Nature . . . is always true . . . she is always right . . . and the mistakes and errors are always those of man." The hypothesis that different pattern genes such as achaete and scute express themselves by means of different prepatterns had to be abandoned. One and the same prepattern seems to exist in the discs of different genotypes. Therefore, differences in visible bristle pattern are the outcome of differential response, to the same prepattern, of cells with different genotypes. A certain bristle is present when the cells differentiating it are genetically competent to respond to the singularity of the prepattern at the specific location. The bristle is absent when the cells for genetic reasons lack this competence. Preceding the determination of 11 pairs of cells toward the developmental path of bristle formation on the mesothorax, a mesothoracic disc in Drosophila possesses eleven pairs of locations, each of which, independently of the others, gives a qualitatively unique signal for bristle production. These different qualities of the 11 different sites will be referred to as singularities. Different alleles endow a cell with specific properties which may be tuned to all these signals as in wild type, or to one or another assortment of them as in the mutants.

It may be noted that the use of the term "competence" in the present context implies an extension beyond its original definition. The term was introduced by Waddington (1932) to signify the epigenetically varying capacity of tissues to respond to inducing stimuli in nonmosaic developing systems. The present extension applies the term to genetically controlled capacities that exist independently of epigenetically acquired ones.

Cryptic Singularities in Prepatterns. The prepattern of the mesothoracic disc of *Drosophila melanogaster* that is involved in bristle formation has more than the 11 singularities that find their expression in wild-type flies. Other related species of flies possess a pair of bristles, the interalars, which normally do not occur in *melanogaster*. Nevertheless, this species too is able to form interalars, provided some specific chromosome aberrations consisting of a duplication are present. Experiments in which discs mosaic for presence and absence of the extra chromosomal material were produced show that the prepattern invariantly includes a singularity at the place of the interalar bristle, and that its nonformation in normal flies is due to lack of response to the singularity of wild-type cells, in contrast to its formation by competent response in the genetically abnormal cells. Such presence of potencies for differentiation that remain latent under usual circumstances is not unique. Very likely another example concerns a certain pair of bristles on the head of *D. subobscura*. Normal flies of the Drosophilidae lack these bristles, which, however, are typical for the closely related family Aulacigastridae. When selection was practiced for an increase in number of bristles and ocelli in a mutant strain of *D. subobscura,* a new pair of bristles appeared, homologous to those in the other family (Sondhi 1962a). Further evidence for the cryptic existence of localized, bristle-inducing singularities in *D. melanogaster* had been provided earlier by intraspecific transplantation experiments carried out by Loosli (1959). The dorsal metathoracic disc of flies of the family Sepsidae forms a group of bristles in a certain region for which there are no homologues in the Drosophilidae. Yet the possibility of formation of these bristles exists in Drosophila, because it can be brought to realization when metathoracic discs are transplanted from one larva to another. The discs then differentiate "adventive" bristles that are homologous to the metathoracic bristles of the distant Sepsidae.

Regional Properties of Prepatterns. The occurrence of adventive bristles is a regional phenomenon rather than a narrowly restricted formation of single bristles at specific positions. Experiments with mosaics include the exploration of regional properties of the prepattern. Thus, the mutant gene hairy-wing (*Hw*), which actually corresponds to a duplication of a single band in the salivary-gland chromosome, places a variable number of extra bristles within a whole area of the thorax that

FIG. 43. Mosaic thorax of *Drosophila melanogaster*. Background heterozygous for the dominant mutant Tufted (*Tft*). Presence of *Tft/+* is marked by black (not-yellow) bristles. An outlined area on the right half of the thorax carries yellow bristles indicating a *+/+* genotype. Autonomously this area has differentiated the normal, non-Tufted pattern of bristles. (From Arnheim 1967.)

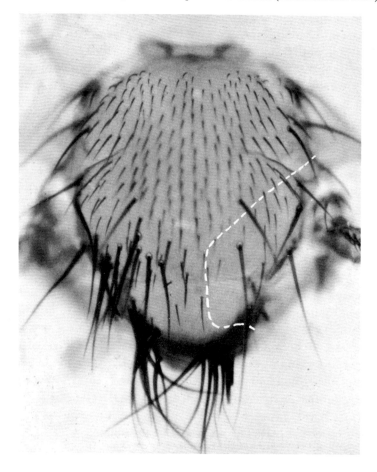

ordinarily exhibits only two pairs of constantly occurring bristles. Mosaic experiments involving *Hw* and not-*Hw* tissues show autonomy in the formation or lack of formation of extra bristles. This indicates an invariant prepattern peculiarity to which *Hw* tissue can respond and not-*Hw* tissue cannot (Gottlieb 1964). Even more striking is the regional effect of the mutant gene Tufted. Tufted flies have a heavy brush of bristles in specific regions of the thorax—without having such tufts elsewhere. Again, mosaics demonstrate the existence of a prepatterned region (Fig. 43; Arnheim 1967). It is noteworthy that the prepatterned regions for the extra bristles of Hairy-wing and Tufted overlap. Apparently the special qualities of the two regions are of different type, and the area of overlap is different from the rest of the thorax in at least two ways.

A mutant that acts analogously to Tufted apparently occurred in one of Hadorn's cultures of genital disc tissues (1966). The normal anal plates of the male have a certain pattern of anterior differentiation of many crowded bristles and of posterior differentiation of loosely distributed bristles. In one of his transplantation lines all bristles were arranged at the same distance from each other. Moreover, instead of forming separate anal plates, the tissue had lost the capacity of arealization and formed large, continuous structures. It is regrettable that this presumed mutant existed in cell cultures only and cannot be subjected to genetic analysis.

Prepattern and Developmental Fields. Operationally the designation "prepattern" for the nonhomogeneities of a developing system provides only limited guidance. While it implies a need to search for differentnesses, it does not suggest any specific type of local properties, either biochemical or physical. Moreover, it includes differentnesses that are not expressed in known later patterns and makes use of multiple differentnesses which distinguish specified regions from others. This weakness of operational guidance should not be taken as a lack of conceptual validity. The dichotomy "prepattern as a term for regional differences existing independently of and preceding the establishment of a subsequent pattern" and "realized pattern in consequence of localized response to the prepattern" seems to me a meaningful device for our understanding, and for the discovery of the existence of singularities at experimentally well-defined locations.

The term prepattern is wider than the term embryonic field. It includes any type of differentness within a developing area, not just those types that have field properties as defined in terms of regulation. Implicitly the singularities of a prepattern are not necessarily interrelated in a way that binds together the components of a field. "Differentness," if defined as prepattern, also is a wider concept than "differentiation," since the prepatterned differentnesses may or may not lead to differentiation.

The realized bristle patterns of mosaic discs in Drosophila provide evidence not only for the existence of invariant overall prepatterns, but also for regulative properties of parts of them. While, for the sake of clarity, autonomy of bristle differentiation has been dealt with exclusively in the preceding discussion, there is evidence for interrelations of bristle differentiation that have field character. Some striking examples are provided by mosaics in the region of the two dorsocentral bristles on each side of the thorax. In wild-type flies these bristles are present in an anterior-posterior line, a certain distance apart. In a mosaic in which the position typical for differentiation of the *posterior* bristle is occupied by achaete (*ac*) tissue, this bristle is not produced in consequence of an autonomous effect of the achaete genotype. If in such a mosaic the position of the typical *anterior* bristle is occupied by cells heterozygous for the dominant normal allele of achaete (*ac/+*), one might expect an anterior bristle to form because this is the case in a nonmosaic fly of *ac/+* constitution.[1] In reality, no bristle is found at its typical position. Instead, a dorsocentral bristle exists at a place *between* the positions of the normal anterior and posterior bristles, a place which normally has no bristle at all (Fig. 44A). In another mosaic thorax the positions of both anterior and posterior bristles were occupied by achaete cells which, autonomously, do not produce either bristle. Instead, a dorsocentral bristle was found *posterior* to the typical position of a posterior bristle (Fig. 44B). Such mosaics demonstrate that the dorsocentral region has a unit character which determines that production of a bristle at a specific location within the narrow dorsocentral line

---

1. Here, and later, the symbol + designates the normal allele of the mutant involved.

suppresses production of bristles close to it. When, however, as in the two examples just described, formation of a bristle at its normal position does not occur, then production of a bristle at a new location is possible. This hypothesis of a suppressive action of bristles on further bristle differentiation in their neighborhood is in line with a corresponding hypothesis on the spacing of bristles in the bug Rhodnius (Wigglesworth 1940, 1959).

The slight modification of autonomous bristle differentiation in mosaic discs due to the existence of restricted field properties is not the only limitation of autonomy. Spread of gene-dependent products from cells of one genotype to cells of another may also lead to bristle differentiation at abnormal positions (Stern 1956). This type of nonautonomy which is superimposed on the essential pattern will not concern us further. Still another type of nonautonomy in the differentiation of mosaics has been discovered. The recessive gene dumpy (*dp*) leads, among other effects, to the appearance of depressions at specific places in the chitinous integument of Drosophila and a whorl-like arrangement of the microchaetae in the depressed and the surrounding regions. In mosaics with patches of homozygous *dp/dp* tissue present, the relevant areas on the phenotypically normal *dp/+* background show the depressions and whorls characteristic of dumpy, but the effect of the *dp/dp* genotype

---

Fig. 44. Two mosaic thoraxes of *Drosophila melanogaster*. White areas, genotype *ac/+* with two dorsocentral bristles (dc) at normal sites. Black areas, genotype *ac*. (A) The *ac* area covers the site of the posterior dc, which therefore does not form. In the *ac/+* area, a dc has formed as expected but between the sites typical for normal dc's. (B) The *ac* area covers the sites of both anterior and posterior dc. In accordance with autonomy of expression of *ac*, no dc bristles are formed within the *ac* area. Posterior to the *ac* area, however, a dc bristle has formed at a site which normally does not form such a bristle. (From Stern 1954*a*.)

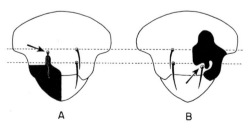

A   B

extends to the neighboring dp/+ tissue so that it participates in vortex-formation (King 1964).

It has been implied in the preceding discussion that at some stage in the development of an imaginal disc its cells are undifferentiated, and that later those cells which find themselves at specific positions of a prepatterned anlage become determined toward bristle formation. Other views of the role of the prepattern are compatible with the main concept. Thus it is conceivable that the invariant prepattern induces the initiation of bristle differentiation even in those locations that in a later stage do not give evidence of it. This would mean that the autonomous genic action does not consist in absence of initiation of differentiation but in degeneration of initiated structures. A bristle pattern would then be the result of two processes, patterned initiation and differently patterned degeneration. Histological studies of imaginal discs of normal and of scute genotypes do not lend support to such a hypothesis (Jacobson 1964).

Another hypothesis would be that certain cells become determined in a nonpatterned way to develop into bristle organs, at locations which do not correspond to those of the later pattern. Rather, the determined cells would migrate within the disc and develop further only if they reached certain places of a prepattern. If this were true, mosaic thoraces should often possess bristles of one marker phenotype (for example, yellow) within an area of another phenotype (for instance, dark) or vice versa. The fact that the phenotypes of bristles in mosaic discs usually are the same as those of their surrounding or neighboring tissue speaks against the migration hypothesis. Even if it were true, the prepattern concept would have to be invoked in the interpretation of the realized pattern.

The Sex Comb of Drosophila Males. The analyses of bristle patterns on the thorax and head of flies have been extended to those of a secondary sex character in normal and mutant strains. The foreleg of the Drosophila male possesses a peculiar structure on its first tarsal segment, its basitarsus. This structure consists of a longitudinal row of approximately 10 to 13 unusually heavy and blunt macrochaetae closely set together. The whole is called the sex comb and the macrochaetae its teeth. The presence of a sex comb is a regional pattern phenomenon which is under genetic control. The sex comb is differentiated at one specific region of the total surface only, a restricted, distal part of the basitarsus, and this differentiation occurs only in individuals of male genotype. Accordingly, the basic question was whether male and female forelegs have essentially different prepatterns or whether male and female cells respond differently to a prepattern invariantly present in both sexes.

Before discussing the results of relevant studies on mosaics, it will be well to give a more refined description of the sex difference of the basitarsus. The ventral surface of this segment presents a series of transverse rows of typical bristles, usually seven or eight in the female and five or six in the male (Fig. 45). The longitudinally arranged sex comb is remarkable on account of not only the size and shape but also the position of its teeth. While the bristles of the transverse rows all point in a distal direction, the teeth point more or less across the basitarsus. The various differences between the two sexes in the appearance of the ventral side of the basitarsus are the results of alternative development of homologous structures.

FIG. 45. The female and male basitarsus and distal part of tibia of *Drosophila melanogaster*. (A) Ventral view of a female basitarsus with seven transverse rows. (D) Ventral view of male basitarsus with six transverse rows and the sex comb. (B), (C) Diagrams of the segments. The cylindrical surfaces are represented as unrolled and projected on a plane. The longitudinal bristle rows are numbered from 1 to 7. The bristle marked ⓒ is the central bristle, which occurs in the male only.

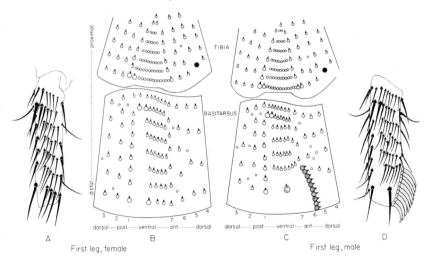

In the female, the basitarsal region distal to approximately the first six transverse rows forms two further transverse rows, while in the male the distal region presumably forms only one row. This row and its anteriorly located adjacent bristles differentiate into the sex comb. Because in the imaginal disc of the male foreleg the prospective sex comb is first determined as a transverse row, its prospective macrochaetae, like those of all other transverse rows, are determined to grow out perpendicular to the direction of the row and thus point distally. But now a shift of tissues occurs that brings the row to a longitudinal position, its initially posterior part coming to lie proximally on the basitarsus and its initially anterior part distally. Automatically and passively the shift of the row causes the change in direction of the teeth which grow out from it.

This fascinating interpretation of the origin of the sex comb has been derived from cell-lineage studies that made use of mosaic basitarsi (Tokunaga 1962). Gynandric basitarsi furnished some of the evidence, but wholly male basitarsi mosaic for pigmentation phenotypes yielded the clearest results. These mosaics had a yellow area on a background of not-yellow, the difference in coloration due to somatic crossing-over in a pair of suitably constituted autosomes. Mosaics were found in which a longitudinal strip of tissue including bristles anterior to the transverse rows was yellow. In these basitarsi yellow teeth were

---

FIG. 46. Three male basitarsi of *Drosophila melanogaster* mosaic for yellow coloration. Different areas of the general surface and of the sex combs are yellow. Solid circles, sockets of black bristles; open circles, sockets of yellow bristles. (From Tokunaga 1962.)

present in the distal part of the comb (Fig. 46A). Other mosaics had yellow bristles along a strip which included the middle of the transverse rows. Here the middle teeth were yellow (Fig. 46B). Still other mosaics had yellow in a strip including posterior bristles of transverse rows. In these the proximal teeth were yellow (Fig. 46C).

The sex difference of the basitarsus is thus accounted for by three processes: broadening of a transverse row, rotational shifting of the row, and differentiation of teeth instead of bristles (Fig. 47). Perhaps the rotation of the row is a consequence of its widening, but the differentiation of teeth is an independent process. This is apparent from the fact that in gynanders single teeth may be formed without shift in their positions, and in certain diploid intersexes a more or less longitudinally arranged row of bristles is formed which consists of enlarged but typical macrochaetae and not of teeth (e.g. Fung and Gowen 1957; Hildreth 1965).

Gynandric Mosaics of the Sex Comb. The problem of the genetic determination of the pattern of male and female basitarsi was studied by means of gynandric mosaics. As in the bristle patterns of the thorax and head in mosaics for achaete and scute genotypes, determination of sex-comb teeth or of bristles occurred autonomously according to the sexual geno-

---

FIG. 47. Cell lineage of sex comb of *Drosophila melanogaster*. t-p, t-m, t-a: posterior, middle, and anterior region of transverse rows. S-p, S-m, S-d: proximal, middle, and distal region of sex comb. Corresponding bristles and teeth are indicated by like markings. (From Tokunaga 1962.)

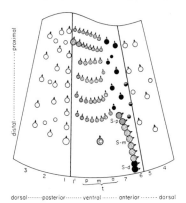

type of the tissue (Stern and Hannah 1950). Whenever the marker gene yellow indicated the presence of male tissue in the region of potential sex-comb formation, one or more teeth were differentiated (Fig. 48*A*); and whenever the marker not-yellow indicated the presence of female tissue, normal bristles were formed (Fig. 48*B*). Particularly striking was the appearance of those mosaics in which the critical region of the basitarsus was mostly made up of tissue of one sex and thus contained only a small area of the opposite sex. When the major part was male and formed a sex comb with a tongue of female tissue extending into the sex comb, then the comb showed a gap in its row of teeth, a gap often occupied by a bristle (Fig. 48*C*). Conversely, when most of the potential sex-comb region was female and formed a transverse row of bristles but had a tongue of male tissue extending into it, then one or more typical teeth appeared as part of the row (Fig. 48*D*). Clearly, formation or nonformation of a tooth depends on the competence of male tissue or lack of competence of female tissue to respond to a prepattern for sex-comb formation that is present in both sexes. Equally, the widening of the transverse row is an autonomous

---

FIG. 48. Gynandric basitarsi of *Drosophila melanogaster*. Dark, female chaetae; light, male chaetae. (*A*) Uninterrupted sex comb with six teeth. (*B*) Distal transverse row of five female bristles. Two sex-comb teeth in abnormal position. (*C*) Sex comb with nine teeth interrupted by one female bristle. (*D*) Distal transverse row of six female bristles. One sex-comb tooth in abnormal position. (From Stern and Hannah 1950.)

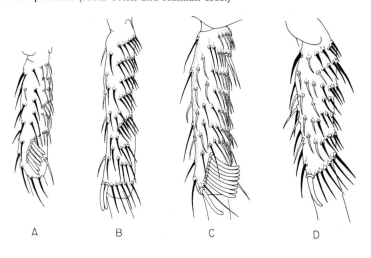

A   B   C   D

response of the male genotype as is apparent from a comparison of the numbers of female bristles and male teeth in mosaics. The number of bristles remains small and corresponds maximally to that characteristic for the last transverse row of a fully female segment. The number of teeth in a mixed transverse row, on the other hand, may greatly exceed the number of corresponding bristles on a female segment. Interestingly, however, there is one nonautonomous feature in certain mosaic basitarsi. This is the position of teeth in those mosaics in which only one or very few teeth are surrounded by female tissue. These few teeth retain their position as part of a transverse row and point distally instead of across the basitarsus. In other words, the shift of position occurs only when a broad section of the transverse row is of male genotype. This is as expected if the shift depends on a widening of tissue as it occurs in a male row.

The Mutant Sexcombless. An interesting sex-linked mutant gene is known which reduces greatly the number of teeth formed on a male foreleg even though its name, "sexcombless" (sx), gives an exaggerated impression of its effect (Fig. 49). Actually, this effect varies considerably from male to male (Mukherjee 1965). Sometimes no teeth are formed, sometimes a comb with a small number of teeth. The teeth themselves may be typically blunt but often are somewhat pointed, and bristles are observed frequently that are in various ways intermediates between macrochaetae and teeth. Gynandric mosaic experiments show autonomy of expression of sexcombless in male

---

FIG. 49. Male basitarsus of a "sexcombless" mutant of *Drosophila melanogaster*. A single tooth and an intermediary bristle are differentiated. (From Stern and Mukherjee 1964.)

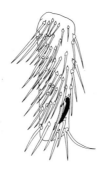

parts, an autonomy that results in the same range of variability as in nonmosaic sx flies, from complete absence of bristle structures other than normal macrochaetae through intermediate forms to typical teeth (Mukherjee and Stern 1965). These results show that a prepattern for sex-comb differentiation is present in sexcombless, but that the reactivity of male tissue is greatly reduced below that of normal males. In other words, the competence of sx tissue lies between that of normal male tissue and that of the noncompetent female tissue.

The Secondary Sex Comb of Engrailed. All our work reported up to this point revealed the presence of invariant prepatterns. There seemed to be no reason why some mutant pattern genes might not produce their effect by changing a prepattern rather than changing the competence of cells. This led to continued search for pattern genes and their analysis by the mosaic method. Having studied the genotypes of normal females and sexcombless males, which lead to differentiation of less than the normal sexcomb, we now turned to genotypes which produce more than the normal sexcomb.

The difference between normal male and female basitarsi consists in the presence of a structure in one sex and its absence in the other. A similar difference within a single sex, the male, is caused by the gene engrailed (en) located in chromosome 2. This recessive mutant allele causes the formation of a secondary sex comb arranged on the basitarsus in mirror image fashion

FIG. 50. Distal halves of three basitarsi of *Drosophila melanogaster.* (A) Homozygous engrailed (en/en) with secondary sex comb of four teeth. (B), (C) Mosaics consisting of phenotypically normal heterozygous en/+ background (dark) and en/en tissues (light) in the region of the secondary sex comb. Seven teeth (B) and one tooth (C) of a secondary comb have formed. (From Tokunaga 1961.)

A   B   C

relative to the typical, primary comb (Fig. 50A). Secondary sex combs of this kind are not found in any other species of Drosophila, and it was hoped that their appearance in engrailed flies was indicative of a basic reorganization of the developing leg disc, that is, the appearance of a new kind of prepattern. Since engrailed is an autosomal gene, mosaics had to be obtained that were not the consequence of chromosome loss but of somatic crossing-over which would cause the production of en/en sections on an en/ + background. As with the formerly discussed gynandric mosaics, it was necessary to make use of a marker gene that would differentiate the engrailed patches from the not-engrailed background independent of the engrailed phenotype. The best marker for distinguishing mosaic patches is the dominant X-linked gene y^+ for not-yellow, normal, dark pigmentation and its allele y for yellow. Using the tricks of Drosophila chromosomal manipulation, in which chromosome breakages are induced by X-rays and desired rearrangements are selected out of the multitude of aberrations, males were constructed which carried y translocated from the X- into the Y-chromosome and, in addition, a not-yellow allele which had been translocated to a second chromosome (Fig. 51). At the same time the males were heterozygous for engrailed, the normal allele being present in the y^+ carrying right arm of the second chromosome and the mutant allele in the homologous chromosome. Phenotypically these males were not-yellow and not-engrailed. When they were subjected to X-irradiation during their larval stage, somatic crossing-over was induced in some of their cells, resulting subsequently in two daughter cells, one of which

---

FIG. 51. Diagram of first and second chromosomes of *Drosophila melanogaster* males heterozygous for engrailed. (*I*) *Above,* X-chromosome deficient for its own tip but with translocated tip of right end of chromosome. *Below,* Y-chromosome with insertion of yellow-carrying X-chromosomal section. (*II*) *Above,* chromosome 2 with not-engrailed (+), deficient for its own right-end tip but with translocated tip of X-chromosome carrying not-yellow. *Below,* homologous chromosome 2 with engrailed. (Redrawn from Tokunaga 1961.)

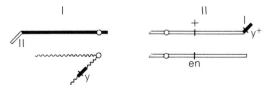

was homozygous for engrailed and free from the y^+ translocation. Multiplication of this cell would lead to the appearance
of a yellow patch which, when it happened to be located in
the area of a potential secondary sex comb, would yield the
desired information on differentiation or nondifferentiation of
a secondary comb.

When the mosaics were studied (Fig. 50B, C), automomy in
production of the secondary comb was found (Tokunaga 1961).
This signifies that even an overwhelmingly phenotypically
normal en/ + male basitarsus possesses a prepattern that singles
out a special area for secondary comb formation, provided the
competent en/en genotype is present. Thus, a singularity in the
prepattern of an en/en basitarsus that controls the formation
of the secondary sex comb is also present in the prepattern of
a phenotypically normal en/ + basitarsus. We conclude that the
engrailed gene fails to qualify as a prepattern mutant. Strictly
speaking, this conclusion goes beyond the evidence as it is
possible that the singularity in an engrailed prepattern has a
character different from that in a not-engrailed prepattern. The
essential fact, however, shown by the experimental results, is
that if the singularities are different they act alike in evoking
formation of a secondary comb by en/en tissue. At present no
means are available to distinguish between the existence of a
single and of two kinds of singularities.

Our hope to find a prepattern mutant was greatly bolstered
when another gene effect was analyzed: the appearance of
primary sex combs on the middle and hind legs of Drosophila
males. It was at first supposed that apart from bearing sex
combs these mutant second and third legs had a total morphology similar to that of typical nonmutant legs, but this was
disproved by a detailed morphological study of the "extra sex
comb" types (Hannah-Alava 1958). Normally, the three pairs of
legs differ from one another in characteristic ways. In addition
to the fact that the male forelegs carry a sex comb and the other
legs do not, there exist differences in the presence or absence
of transverse rows on various segments of the three pairs of
legs and in the regions where such rows occur. Still other
differences concern the presence or absence of unique large
bristles. The comparative morphological studies of Hannah-
Alava established the remarkable fact that the second

and third legs of extra-sex-comb mutants were not typical legs of their kinds with an added sex comb. Rather, they were legs whose ventral morphology made them similar to first legs. The extra-sex-comb mutants are thus homoeotic mutants that change certain organs, the middle and hind legs, to resemble—at least in part—the forelegs. In detail there is much variation from one individual to the next (Fig. 52). Thus the distal end of the tibia and the basitarsus of a second leg may be fully like those of a normal second leg without transverse rows and without sex-comb teeth. Or it may have one or two teeth on an otherwise typical second leg, or transverse rows on the tibia or the basitarsus as on a first leg and many teeth of a typical sex comb. Strangely enough, in spite of a statistical correlation between the degrees of "first-leggedness" of different parts, a variety of morphological combinations showed that homoeotic transformation of one part of a leg is independent of other parts. Each part, in a developmentally mosaic way, is able to change toward features of a foreleg.

The homoeotic nature of the extra-sex-comb mutants pro-

FIG. 52. Different degrees of genetic transformation of basitarsus and part of tibia of second legs of *Drosophila melanogaster* into first legs. Note sex combs and transverse rows on basitarsus and tibia. (From Hannah-Alava 1958.)

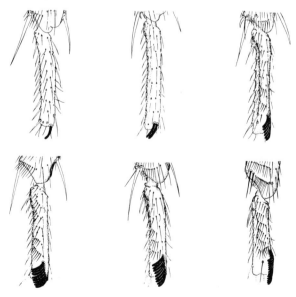

vided a suggestive answer to the question why these mutants call forth sex combs on second and third legs. Because the mutants change these legs more or less into first legs, and because male first legs typically have sex combs, the appearance of the extra combs seemed a secondary consequence of a primary homoeotic transformation. The mutant genes seemed to produce their effect by changing the basic organizations of the second and third pair of limb discs into one approaching the basic organization of the first disc. The presumed likelihood that the different patterns of the normal three pairs of legs were built upon three different prepatterns led to the expectation of nonautonomy of the features of extra-sex-comb second and third legs in genetic mosaics. Because a second leg heterozygous for a recessive extra-sex-comb gene (*esc*/ +) has the typical

---

Fig. 53. Autonomy in differentiation of mosaics for extra sex comb in *Drosophila melanogaster*. *II*, basitarsus of second leg; *III*, sections of third leg. Background *esc*/ + normal, dark. Areas of *esc*/*esc*, light. (A) A very small basitarsal *esc*/*esc* patch differentiates a sex-comb tooth. (B) A patch differentiates several bristles and two teeth. (C) A patch differentiates several bristles including a transverse tibial row. (From Tokunaga and Stern 1965.)

A

B

C

features of a normal second leg, a small patch of homozygous esc/esc tissue was thought not able to find a suitable prepattern to let it develop structures characteristic for a first leg. Rather it was expected that esc/esc tissues would behave nonautonomously by nonformation of a sex comb and other first leg features.

The required mosaics were obtained by induction of somatic crossing-over using a method similar to that designed for engrailed, but adjusted to the fact that esc is located on the left arm of chromosome 2. Again, however, the expectation of nonautonomy due to lack of a suitable prepattern was disappointed. When the mosaic specimens were studied, it was apparent that whenever esc/esc tissue occurred in a critical region on a second or third leg, this tissue was inclined to form appropriate parts of transverse rows or sex-comb teeth, the inclination being expressed in a fraction of mosaics only, corresponding to the incomplete penetrance of the effect of esc/esc in whole flies (Fig. 53; Tokunaga and Stern 1965). Thus the three pairs of legs are alike in having the same prepattern to which different genotypes, extra-sex-comb and normal, respond differentially. All legs have a singularity for potential comb formation and all are able to cause the appearance of transverse rows of the kind normally found on first legs only.

Fig. 54. Antennal mosaics in *Drosophila melanogaster*. Background heterozygous for the recessive mutant aristapedia (ss^a/+), dark bristles and hairs. Homozygous ss^a/ss^a tissue, bristles and hairs in outline. (A) Frontal view of a head which bears a normal antenna with feather-like arista (right) and a mosaic antenna with tarsal segments (left). (B) Antenna in which wild-type tissue autonomously forms aristal filaments though surrounded by ss^a/ss^a tissue. (C) Distal part of an antenna in which tissues of the two genotypes are intricately intermingled. (From Roberts 1964.)

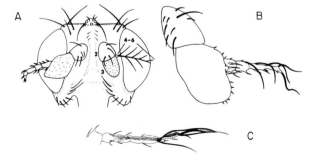

Autonomy of differentiation of homoeotic mutant tissues in mosaics has also been described in Drosophila for the bithorax alleles which transform certain segments of the thorax and abdomen into serially homologous segments (Lewis 1963), as well as for the mutant aristapedia which leads to formation of a tarsus in place of an antennal segment (Fig. 54; Roberts 1964). For either case, and as in the extra-sex-comb mutants, the picture is that of invariant prepatterns but differential reactivity to them.

Epigenetic Differences in Prepatterns. What are the underlying differences in development that lead to morphological differences between the three pairs of normal legs in a genetically uniform, normal individual? A general answer to this question may be based on one or the other of two not mutually exclusive hypotheses. In hypothesis (A) it is assumed that the singularities for the evocation of the specific features of a first leg follow an anterior-posterior gradient of decreasing "strength," and that the reaction of cells with the normal genotype requires a minimum level of evocation which is reached or surpassed in the first leg disc only (Fig. 55 A). For the extra-sex-comb genotypes the reaction toward development of a first leg has a lower threshold than for normal, so that the minimum strength of evocation is adequate, even in the second and third leg discs, to lead to homoeotic transformation. Alternatively in hypothesis (B) it is assumed that the strength of evocation is the same for all leg discs, but that the threshold for response toward differentiation of a first leg increases along the anterior-posterior axis (Fig. 55B). As a result, cells with the normal genotype are unable to form homoeotic first legs by the second and third discs. Cells with extra-sex-comb genotypes with their lower thresholds of response would still be able at the positions of the second and third discs to enter homoeotic transformations.

Either hypothesis implies epigenetically acquired differences of genetically identical cells in the three anlagen. Either hypothesis introduces a third notion into the scheme of pattern formation as deduced from mosaic experiments. There is added to the concepts of prepattern and genetic competence that of epigenetically varying strength of a prepattern or of epigenetically varying competence. These latter concepts may be regarded as subordinate to those of prepattern and competence since they serve only to specify variable aspects of the two main features.

The existence of epigenetically acquired different properties in serially homologous organs had been earlier demonstrated in the moth Ephestia (Kroeger 1959). This investigator succeeded in combining early imaginal discs of fore- and hindwings and culturing them as implants into larval hosts. The combined tissues formed a unified prepattern for wing hinges as was deduced from the joint formation of hinge complexes by the original fore- and hindwing tissues. However, the forewing tissue differentiated in the direction of forewing hinge parts and the hindwing tissue toward hindwing hinge parts.

The Case for Prepattern Mutants. It is pleasant to find one's predictions come true, but there is often even more reason to be elated when the opposite happens. An unexpected result may

FIG. 55. Alternative models of the prepatterns of the legs of *Drosophila melanogaster* and the competence threshholds for *esc⁺* and *esc*. (A) Decreasing strength of singularity, unchanged competence for the three legs. (B) Equal strength of singularity, increasing threshholds. (From Tokunaga and Stern 1965.)

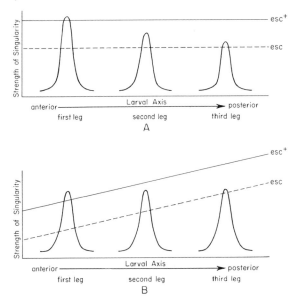

lead to new insights. The unexpected autonomy of bristle differentiation in mosaics for achaete and scute led to the concepts of prepattern and genetic control of competence. The autonomy of the second sex comb in engrailed revealed an unknown singularity in the forelegs of Drosophila. The autonomy of differentiation of the homoeotic mutants showed that the determination of the features characteristic for each different leg, such as transverse rows and sex combs, or of antennal versus leg features occurs separately for each of these properties. The decision on what is to be realized out of a variety of alternatives is made at each specific region; there is no overall organization of an imaginal disc that would tend to decide on the development of the type of appendage as a whole.

Notwithstanding these positive aspects of the work, it became an urgent question at this stage of our investigations why the search for causal genetic factors in the initiation of the terminal bristle and sex-comb patterns as well as other patterns in Drosophila had revealed only differences in competence of tissues but not differences in their prepatterns. Development may be conceived as a consecutive series of patterned events in which a specific prepattern becomes expressed in a subsequent pattern, the subsequent pattern serving as a prepattern for the next patterned processes, and so on until a late prepattern forms the basis of a terminal pattern. One would expect that mutant genes exist that control variations of either an early prepattern or of the competence of cells to respond to an early prepattern, and that in both cases the subsequent pattern would differ from the normal one and serve as an altered prepattern for the next step. The fact that none of the observed variations in terminal patterns had been traced to variations in the immediately preceding prepattern seemed to signify either that the constancy of the prepattern is genetically strongly guarded and that a deviation may be lethal, or that deviations are regulatively compensated so that invariance is secured. These possible phenomena for which Waddington (1957 and various earlier discussions) uses the picture of "canalisation of development" remain obscure.

There is no basic difficulty in conceiving changes in prepattern. This may be illustrated by a hypothetical example. The

Himalayan rabbit exhibits an interesting series of patterns of pigmentation. The best-known type is the rabbit pigmented at the snout, ears, tail, and feet but more or less unpigmented elsewhere (Fig. 56). This pattern of pigmentation is based on a prepattern of temperature differences in the skin which presumably occur in all breeds of rabbits (review in Danneel 1941). The temperature differences are partly due to differences in the amount of fur, but probably they mostly result from an insufficiency of blood vessels to carry warm blood to the body surface and thus maintain terminal parts of the animal at the same temperature as the main parts of the body. Rabbits which carry the gene C for normal pigmentation are uniformly colored, since in such a genotype the production of melanin proceeds both at low and at high temperatures. Rabbits which carry the allele c in homozygous constitution are uniformly white because in cc cells the production of melanin does not occur at either temperature. Rabbits, however, that carry the genotype $c^h c^h$ have melanocytes that produce pigment at low but not at high temperatures. A critical skin temperature exists above which no pigment is formed. It is alike for all body regions and lies at about 33° C. The actual skin temperature varies from one part of the body to another. Depending on the outside temperature, different body regions attain the critical temperatures selectively. At outside temperatures higher than 29° all

---

FIG. 56. Himalayan rabbit (Moscow strain) raised at a temperature somewhat below 14°C. The temperatures listed at different parts of the body represent those critical values of the outside temperature above which no pigment is formed at that body region. (Redrawn from Iljin 1927.)

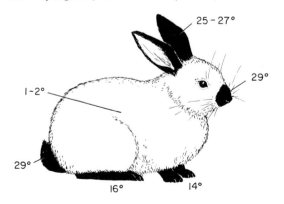

parts of the skin would be above 33° and the whole animal would be white. At an outside temperature below 1–2° no part of the skin would be above 33° and the whole animal would be colored. At intermediate outside temperatures some areas of the skin would be above and others below the critical value. For instance, at 24° outside only the snout, ears, and tail would be below the critical temperature and thus be pigmented; at 15° the hindfeet, in addition to the snout, ears, and tail, would be colored, and only at temperatures somewhat below 14° would the pattern of pigmentation be that shown in the figure. There are thus a whole series of different pigmentation patterns.

The difference between the pattern of pigmentation in the Himalayan rabbit and the uniformity in other genotypes consists of genetically controlled competence differences of the melanocytes. Yet it is not difficult to imagine that selection might produce changes in the distribution and efficiency of blood circulation in the terminal parts which would result in sufficiently warm feet and other "akra" (tips), so that the whole Himalayan animal would remain unpigmented at any temperature. Similarly, selection might lead to higher skin temperatures of some specific areas and thus result at any temperature in one or the other of the specific pigmentation patterns discussed above. The pattern differences between these hypothetical rabbits and the present Himalayan would be based on genetic differences in the temperature prepatterns and not on genetically different competence of melanocytes.

An actual difference in prepatterns has been briefly described for two species of toads. *Bufo regularis* and *Bufo carens* (Balinsky 1956). In tadpoles of these species the number of pairs of external gills differs, being two in *regularis* and three in *carens*. Interspecific transplantation experiments demonstrate that it is the embryonic tissue underlying the ectoderm of the head region that induces the differentiation of external gills. Moreover, it is the underlying tissue which differentially determines the number of gills induced. Epidermis of *regularis* forms three gills when transplanted to *carens* head mesoderm, and epidermis of *carens* on a *regularis* head forms two gills. The ectoderm of either species is competent to respond to the induction stimuli of the other species, but the number of induced structures is determined not by the tissues forming them but, nonautonomously, by the inductor.

Eyeless-dominant as a Prepattern Mutant. What is true for toads finally proved to be valid also for Drosophila. In our search for prepattern differences we studied the morphological expression of the genotype eyeless-dominant (ey^D). It had been reported that this mutant not only influences differentiation of the eyes but in addition causes abnormalities of the legs and the sex comb. The second tarsal joints were "shortened to give lumps" (Bridges and Brehme 1944), and the sex comb showed "a hypertrophy... being doubled in size" (Patterson and Muller 1930). These brief descriptions had not prepared us for what we found. Instead of a general, irregular hypertrophy of a sex comb in the basitarsal region of male forelegs that normally bears a single comb of some 10–13 teeth, there was a whole group of sex combs, often arranged more or less parallel to one another and consisting frequently of as many as 40–50 teeth (Fig. 57A). Further inspection of the forelegs of both male and female ey^D flies suggested an explanation for the origin of the multiple combs. The shortening of the second tarsal segment, and the lumps generated by it, are correlated with a widening of the distal part of the basitarsus and an incomplete separation between it and the second segment. In many flies of both sexes the two segments form a more or less continuous structure (Fig. 57B). There are several incomplete constrictions only one of which is present normally and, if fully formed, articulates the basitarsus with the next segment. On female forelegs the incomplete separation of the two segments and the widening of parts of the first are accompanied by distortions and other changes in the ventral transverse bristle rows of the basitarsus. There is an increase in rows and numbers of bristles in the distal area. Because, as we have seen, during development of the normal male the last transverse row becomes shifted and transformed into the single sex comb, the ey^D situation became intelligible by assuming that the surplus of transverse rows in the ey^D female is equivalent to the surplus of sex combs in the ey^D male. Instead of the last row in not-ey^D flies, the surplus rows also seem to undergo shifting and transformation into sex combs in ey^D. Expressed differently, the incomplete separation

FIG. 57. The effect of the mutant eyeless-dominant (ey^D) on the foreleg of *Drosophila melanogaster*. C, central bristles. (A) Male basitarsus and incompletely separated second tarsal segment with multiple sex comb. (B) Corresponding female segments. (C) Not-yellow/yellow mosaic distal part of male basitarsus. The circles to the right of the drawing show the position of sex-comb teeth and central bristles. (D) Another example of a not-yellow/ yellow mosaic region of a male basitarsus. (A, C, D from Stern and Tokunaga 1967; B original.)

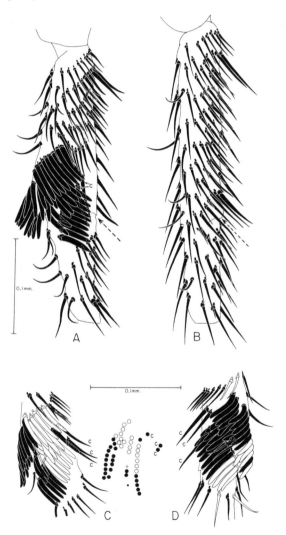

of the tarsal segments established a prepattern which led to excess transverse rows, and the usual differentiation phenomena acting on the terminal rows were now applied to the terminal and to the excess rows.

This interpretation does not require changed competence in ey^D tissues. Instead, it postulates a greatly enlarged area of the prepattern singularity which in the male induces the differentiation of sex-comb structures. This means a change in the prepattern itself. Assuming this hypothesis to be correct, it was predicted that in mosaic forelegs not-ey^D tissue occupying part of the sex-comb area of a predominantly ey^D basitarsus would show nonautonomy in sex-comb differentiation. The predominantly mutant prepattern would induce normal not-ey^D cells to form sex combs at abnormal sites.

This time the prediction was confirmed (Fig. 57C, D; Stern and Tokunaga 1967). The evidence, though, was somewhat less direct than in the autonomous cases discussed earlier. The normal not-ey^D tissue patches, $+/+$, were marked as in other mosaics by yellow coloration, but the variability in location of somatic crossovers along the chromosome is such as to produce two kinds of yellow patches, those that still have the original $ey^D/+$ genotype and those that are $+/+$. The $ey^D/+$ cells, of course, would form surplus sex-comb structures since their genotype would be the same as that of the $ey^D/+$ background. The $+/+$ yellow cells would be distinguishable from the $ey^D/+$ yellow cells if they behaved autonomously and did not form surplus combs. If, however, the $+/+$ cells behave nonautonomously and do form surplus combs, their phenotypic effect would not be distinguishable from that of the $ey^D/+$ cells. The fact is that essentially all yellow patches participated with the background tissue in forming surplus combs although independent tests made it likely that over two-thirds of the patches were $+/+$. Even though it was not possible to point at specific mosaics as being the nonautonomous ones, it was concluded that approximately 33 of the critical total of 47 mosaics belong to the nonautonomous class.

Nonautonomy in mosaics may not only be the consequence of prepattern differences. It can also be brought about by diffusion of gene-dependent substances from one genetic tissue type to another or by predetermination of the activities of cells during the time in which they still had their initial genotype, that is, before they became different as a result of chromosome loss or somatic crossing-over. In the case of ey^D these latter possibilities can not be excluded with certainty, but it seems more likely that the nonautonomy of $+/+$ tissue on an $ey^D/+$ background is to be attributed to a prepattern differential between the normal and the mutant genotype.

There is also a suggestion of an effect on prepattern by a mutant other than ey^D. This is "Hairy wing" (Hw), which was cited earlier as changing the competence of hypodermal cells to respond to a regional singularity. In addition to this action Hw also alters the relative dimensions of the thorax so that the differentiation of microchaetal bristles occurs according to a somewhat changed pattern (Gottlieb 1964). The underlying difference in dimensions may represent a difference between prepatterns. Such mutant effects on a prepattern, including the enlargement of the sex-comb-producing area in ey^D flies, may appear trivial. This adjective, however, can be readily applied to any element in the *Gestalt* of biological processes. Only as part of the organismal unit is an elementary event not trivial.

The findings on patterned differentiation that have formed the material of this presentation may finally be summarized in a generalized diagram (Fig. 58). A basic prepattern I consists

---

FIG. 58. A generalized diagram of prepattern and competence interactions. See text for explanation.

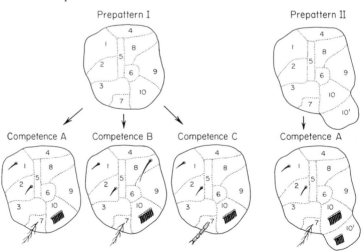

of an assembly of 10 different singularities. Three genotypes, A, B, and C, are shown to respond to prepattern I: A by being competent for specific differentiations at the singularities 1, 2, 7, and 10; B at 1, 2, 7, 8, and 10; and C at 1, 2, 10, and in a new manner, at 7. No response by any one of these genotypes occurs at the singularities 3, 4, 5, 6, and 9. A mutant prepattern II differs from I by having an enlargement (10′) of the area of singularity 10. Genotype A reacts to prepattern II by differentiation of a duplicate structure in area 10–10′.

The Immediate Expression of Changed Genotypes. We have mentioned the possibility that nonautonomy in the expression of a tissue patch could be due to predetermination, a persistence of the influence of the cell-ancestral genotype before the occurrence of the genetic change. Autonomy represents the opposite situation, that of a genetic alteration expressing itself regardless of the prior genotype. In this case the cell from which a chromosome is lost or in which genetic changes occur as the result of somatic crossing-over seems to "forget" the fact that up to the time of the change its genotype was of a different kind than that which it acquired at the moment of change. This is particularly striking when the change occurs at a very late stage of development. Sometimes a single bristle appears to develop from a changed area, and one may wonder whether the moment of change occurred at one of the latest divisions of the cells which then differentiated into the bristle according to its new genotype. That changes can actually occur so late and be expressed immediately is likely from the results of some mosaic experiments in which the gene "multiple wing hairs" (*mwh*) is involved. This recessive mutant expresses itself in the individual cells of the wing and of other surface areas by formation of several cell hairs instead of a single one as is typical under the influence of the normal allele. Wings of flies heterozygous for *mwh*/ + are composed of cells with single hairs but areas occur, probably in consequence of somatic crossing-over resulting in *mwh*/*mwh* constitution, in which each cell bears multiple hairs. An area may even consist of a single cell. It must be noted, however, that the interpretation of genetic mosaicism for these phenotypic mosaics is not yet proven.

Many years ago, the same phenomenon was shown for a mutational type of mosaics. In the plant *Delphinium ajacis* the rose-alpha gene mutates frequently to its allele for purple petal color (Demerec 1931). Depending on the time in development when the mutation takes place, large sectors of a plant may bear purple flowers or small areas of a single petal may be purple (Fig. 59). Some of these areas consist of single cells only, indicating that the mutation occurred at the end of cell division in the petal. The immediate expression of the new genotype either signifies that the genes involved become active only at the very last period or that the products of the preceding genotype have such a short life that they do not enter into the subsequent differentiation. Both phenomena seem to play a role in development. It may be added that a similar conclusion applies to those cases where the immediate products of meiosis exhibit the results of their newly acquired segregational genotypes (Darlington 1937; Stern 1938).

Activation of Pleiotropic Genes. It is tempting to equate the differentiating reaction of cells to the singularities of prepatterns with the calling into action or the de-repression of the genes responsible for pattern differences. For example, the engrailed allele that controls the formation of a secondary sex comb would be thought of as being inactive in most cells of the developing fly except in the region of this comb. It is known, however, that this gene is also involved in morphogenetic processes of other kinds in other regions of the body. Engrailed

---

FIG. 59. The immediate effect of a mutable gene in *Delphinium ajacis*. (A) Flower showing areas of purple color (dark) on a background of rose. Mutations occur from recessive rose to dominant purple. (B) A purple area consisting of four cells surrounded by cells of intermediate coloration. The latter cells are presumably of rose genotype, darkened by diffusion from the purple cells. (C) A spot consisting of a single purple cell with its halo of intermediately colored cells. (Redrawn from Demerc 1931.)

A B C

affects the formation of wing veins and the development of the scutellum. It is possible that these pleiotropic effects are the consequences of a primary effect in some organ that secondarily impinges on basitarsus, wing, and scutellum. It seems more likely that engrailed is involved independently in differentiation at the three sites. One may assume either that the gene is called into action at these sites and not elsewhere, or one may assume that the gene is active in many other cells but that its activity does not express itself outside of the three sites. It may be justified to dwell somewhat on these alternatives. If it were possible to determine cytologically the activity of the engrailed gene in cells of many different sites, one might find that RNA synthesis at the engrailed locus occurs only at the three sites, A, B, and C, where engrailed effects are phenotypically apparent. On the other hand one might find that synthesis at the engrailed locus occurs at more sites than A, B, and C but that in addition site A differs from sites B and C by synthesis at some other locus, x, than en, that site B differs from A and C by synthesis at a locus y, and that site C differs from A and B by snythesis at a locus z. In this scheme the multiple effect of the gene engrailed would depend on its specific interactions with other genes at different sites.

The experiments that have been discussed in this essay and their interpretation may be regarded as the issue of a late marriage between classical experimental embryology and classical genetics. This offspring enters the world of biology at a period when both disciplines have grown to new dimensions.

REFERENCES

Arnheim, N., Jr.
   1967   The regional effects of two mutants in Drosophila analyzed by means of mosaics. *Genetics* 55: 253–263.

Balinsky, B. I.
   1956   Discussion of Dr. Zwilling's paper. *Cold Spring Harbor Symp. Quant. Biol.* 21: 354.

Borstel, R. C. von
   1963   Cytogenetics and developmental genetics course—a postscript. *Am. Zool.* 3: 87–95.

Bridges, C. B., and K. S. Brehme
1944   The Mutants of *Drosophila melanogaster*. *Carnegie Inst. Wash. Publ.* 552. 257 pp.

Claxton, J. H.
1964   The determination of patterns with special reference to that of the central primary skin follicles in sheep. *J. Theoret. Biol.* 7: 302–317.

Danneel, R.
1941   Phänogenetik der Kaninchenfärbung. *Ergeb. Biol.* 18: 55–87.

Darlington, C. D.
1937   Interaction between cell nucleus and cytoplasm. *Nature* 140: 932.

Demerec, M.
1931   Behaviour of two mutable genes of *Delphinium ajacis*. *J. Genet.* 24: 179–193.

Driesch, H.
1894   *Analytische Theorie der organischen Entwicklung*. Leipzig: W. Engelmann. 184 pp.

Dubinin, N. P.
1929   Allelomorphentreppen bei *Drosophila melanogaster*. *Biol. Zent.* 49: 328–339.

Fung, S. T. C., and J. W. Gowen
1957   The developmental effect of a sex-limited gene in *Drosophila melanogaster*. *J. Exptl. Zool.* 134: 515–532.

Gottlieb, F. J.
1964   Genetic control of pattern determination in Drosophila. The action of Hairy-wing. *Genetics* 49: 739–760.

Hadorn, E.
1965   Ausfall der Potenz zur Borstenbildung als "Erbmerkmal" einer Zellkultur von *Drosophila melanogaster*. *Z. Naturforsch.* 20 b: 290–292.
1966   Über eine Änderung musterbestimmender Qualitäten in einer Blastemkultur von *Drosophila melanogaster*. *Rev. Suisse Zool.* 73: 253–266.

Hadorn, E., G. Anders, and H. Ursprung
1959   Kombinate aus teilweise dissoziierten Imaginalscheiben verschiedener Mutanten und Arten von Drosophila. *J. Exptl. Zool.* 142: 159–175.

Hannah-Alava, A.
1958   Developmental genetics of the posterior legs in *Drosophila melanogaster*. *Genetics* 43: 878–905.

Hildreth, P. E.
1965   Doublesex, a recessive gene that transforms both males and females of Drosophila into intersexes. *Genetics* 51: 659–678.

Hollingsworth, M. J.
1964   Sex-combs of intersexes and the arrangement of the chaetae on the legs of Drosophila. *J. Morphol. 115*: 35–52.

Iljin, N. A.
1927   Studies in morphogenetics of animal pigmentation. IV. Analysis of pigment formation by low temperature. (Russian, with English summary.) *Trans. Lab. Exptl. Biol. Zoo park* (Moscow) *3*: 183–200.

Jacobson, C.
1964   The embryology of canalised characters in mice and Drosophila. Ph.D. diss., University of Sydney, Australia.

King, J. L.
1964   The formation of dumpy vortices in mosaics of *Drosophila melanogaster. Genetics 49*: 425–438.

Kroeger, H.
1959   Determinationsmosaike aus kombiniert implantierten Imaginalscheiben von *Ephestia kühniella* Zeller. *Arch. Entwicklungsmech. Organ. 151*: 113–135.

Lewis, E. B.
1963   Genes and developmental pathways. *Am. Zool. 3*: 33–56.

Loosli, R.
1959   Vergleich von Entwicklungspotenzen in normalen, transplantierten und mutierten Halteren-Imaginalscheiben von *Drosophila melanogaster. Develop. Biol. 1*: 24–64.

Maynard Smith, J., and K. C. Sondhi
1960   The genetics of a pattern. *Genetics 45*: 1039–1050.
1961   The arrangement of bristles in Drosophila. *J. Embryol. Exptl. Morphol. 9*: 661–672.

Mukherjee, A. S.
1965   The effects of sexcombless on the forelegs of *Drosophila melanogaster. Genetics 51*: 285–304.

Mukherjee, A. S., and C. Stern
1965   The effect of sexcombless in genetic mosaics of *Drosophila melanogaster. Z. Vererbungslehre 96*: 36–48.

Nöthiger, R.
1964   Differenzierungsleistungen in Kombinaten, hergestellt aus Imaginalscheiben verschiedener Arten, Geschlechter und Körpersegmente von Drosophila. *Arch. Entwicklungsmech. Organ. 155*: 269–301.

Patterson, J. T., and H. J. Muller
1930   Are "progressive" mutations produced by X-rays? *Genetics 15*: 495–578.

Roberts, P.
1961   Bristle formation controlled by the achaete locus in genetic

mosaics of *Drosophila melanogaster*. *Genetics 46*: 1241–1243.

Roberts, P.
1964 Mosaics involving aristapedia, a homeotic mutant of *Drosophila melanogaster*. *Genetics 49*: 593–598.

Scharloo, W., M. S. Hoogmoed, and W. Vreezen
1966 Pattern formation and canalization. *Genen en Phaenen 11*: 1–15.

Serebrovsky, A. S.
1930 Untersuchungen über Treppenallelomorphismus. IV. Transgenation scute-6 und ein Fall des "Nicht-Allelomorphismus" von Gliedern einer Allelomorphenreihe bei *Drosophila melanogaster*. *Arch. Entwicklungsmech. Organ. 122*: 88–104.

Sondhi, K. C.
1962a The evolution of a pattern. *Evolution 16*: 186–191.
1962b Patterns of gene pleiotropy in morphogenetic processes. *Science 137*: 538–540.
1963 The biological foundations of animal patterns. *Quart. Rev. Biol. 38*: 289–327.

Spemann, H.
1936 *Experimentelle Beiträge zu einer Theorie der Entwicklung.* Berlin: J. Springer. 296 pp.
1938 *Embryonic Development and Induction.* New Haven: Yale Univ. Press. 401 pp.

Stern, C.
1938 During which stage in the nuclear cycle do the genes produce their effects in the cytoplasm? *Am. Naturalist 72*: 350–357.
1954a Two or three bristles. *Am. Scientist 42*: 213–247.
1954b Genes and developmental patterns. *Proc. Intern. Congr. Genet. 9th (Bellagio, Italy 1953) Caryologia (suppl.) 6*: 355–369.
1956 The genetic control of developmental competence and morphogenetic tissue interactions in genetic mosaics. *Arch. Entwicklungsmech. Organ. 149*: 1–25.
1965 Entwicklung und die Genetik von Mustern. *Naturwiss. 52*: 357–365.

Stern, C., and A. M. Hannah
1950 The sex combs in gynanders of *Drosophila melanogaster*. *Port. Acta Biol. Ser. A. Vol. R. B. Goldschmidt*: 798–812.

Stern, C., and A. S. Mukherjee
1964 Aspects of the developmental genetics of the legs of *Drosophila*. *Proc. Nat. Acad. Sci. India Sect. B. 34*: 19–26.

Stern, C., and D. L. Swanson
1957 The control of the ocellar bristle by the scute locus in *Drosophila melanogaster*. *J. Fac. Sci. Hokkaido Univ. Ser. VI 13*: 303–307.

Stern, C., and C. Tokunaga
1967   Nonautonomy in differentiation of pattern-determining genes in Drosophila. I. The sex comb of eyeless-dominant. *Proc. Nat. Acad. Sci. U.S. 53:* 658–664.

Steward, F. C.
1964   Growth and development of cultured plant cells. *Science 143:* 20–27.

Sturtevant, A. H.
1932   The use of mosaics in the study of the developmental effect of genes. *Proc. Intern. Congr. Genet. 6th (Ithaca, N.Y. 1932)* 1: 304–307.

Tokunaga, C.
1961   The differentiation of a secondary sex comb under the influence of the gene engrailed in *Drosophila melanogaster. Genetics 46:* 157–176.
1962   Cell lineage and differentiation on the male foreleg of *Drosophila melanogaster. Develop. Biol. 4:* 489–516.

Tokunaga, C., and C. Stern
1965   The developmental autonomy of extra sex combs in *Drosophila melanogaster. Develop. Biol. 11:* 50–81.

Torrey, J. G.
1966   The initiation of organized development in plants. *Advan. Morphogenesis 5:* 39–91.

Turing, A. M.
1952   The chemical basis of morphogenesis. *Phil. Trans. Roy. Soc. London, Ser. B 237:* 37–72.

Ursprung, H.
1963   Development and genetics of patterns. *Am. Zool. 3:* 71–86.
1966   The Formation of Patterns in Development. In *Major Problems in Developmental Biology, 25th Symp. Soc. Devel. Biol.,* ed. M. Locke, pp. 177–216. New York: Acad. Press.

Waddington, C. H.
1932   Experiments on the development of chick and duck embryos, cultivated in vitro. *Phil. Trans. Roy. Soc. London, Ser. B 221:* 179–230.
1957   *The Strategy of the Genes.* New York: Macmillan Co. 262 pp.

Weiss, P.
1939   *Principles of Development.* New York: Henry Holt. 601 pp.

Wigglesworth, V. B.
1940   Local and general factors in the development of "pattern" in *Rhodnius prolixus* (Hemiptera). *J. Exptl. Biol. 17:* 180–200.
1959   *The Control of Growth and Form.* Ithaca: Cornell Univ. Press. 140 pp.

Young, S. S. V., and R. C. Lewontin
1966   Differences in bristle-making abilities in scute and wild-type *Drosophila melanogaster. Genet. Res. 7:* 295–301.

Thoughts on Research

The joys of the investigator have often been sung. The ecstasy which led Archimedes to rush from the bath into the street, naked and shouting "Eureka!" (I have found it!), has been felt by all discoverers even though their restraint in exhibiting their ecstasy made it less memorable.

To perceive a fact of nature that had never been seen before by any human eye or mind, to discover a new truth in any field, to uncover an event of past history or discern a hidden relation, these experiences the fortunate bearer will cherish throughout his life. But let the pains of research not be overlooked. Hardly has the new discovery been made when its author begins to question its validity. How the heart can seem to cease beating when a loophole is perceived! In such a moment the heaven-high jubilance of the discoverer turns into deathly sadness. And it is not sickly indecision which brings this sudden despair; rather, it is the investigator's task to doubt and to doubt again.

Many supposed discoveries wither under the impact of such criticism. Some, however, will finally withstand all tests. Now new questions arise. What is to be done next? Can one pursue further the lead which brought success, or has success itself closed the path to continued harvest in its area? Should one start somewhere else, and what are the chances that important insights will be gained in the new field?

"Knowledge is like a sphere in space," wrote Pascal, "the greater its volume, the larger its contact with the unknown." This may indeed be true but contact with the unknown is not enough. It is of the essence not just to ask questions but to ask the right ones. There is rarely an answer to wrong or vague questions. No biologist has approached his subject successfully when he began by asking "What is life?" He did succeed in a more modest way when he posed such questions as "What is the chemistry of fermentation?" or "How does water ascend a tree?" New discoveries can be the end as well as the beginning of a period of research. Pascal's beautiful analogy has an ugly counterpart in another analogy, that of a piece of wood inhabited by termites, the wood representing the problem and the insects the researchers. When the termites have reduced the log to dust the problem is solved—and the search for another log may be long. Thus many physicists of the late nineteenth century felt that they had essentially solved the problems of their science and that little remained to be done. It is a happy thought that unexpectedly they became challenged again by the discoveries of radioactivity and X-rays.

Knowledge grows by the rare findings of great importance as well as by the accretion of minor ones. Unfortunately many minor findings do not simply add up to one major. Major and minor are labels which derive not from the facts or concepts themselves, but from their significance to the whole. Theodor Boveri, the great biologist, once wrote as follows: "The significance of a discovery is determined much less by its specific achievement than by subsequent investigations which show whether the validity of the findings is narrow or wide or even all-embracing."

This significance cannot be foreseen. Here the course of science resembles that of evolution. It may be pictured as an exploration of an unending series of mountain chains. When you enter a new valley you cannot know whether it will end blindly or lead to a pass through which one may reach a vast new area. There are few passes and many dead ends. Many species become extinct without having evolved into new ones. Those in existence now are the descendents of the few who happened to cross the barriers. Man himself had a very narrow escape from never coming into existence. If it is true that most explorations by investigators and species do not lie in the line which leads to the future, it follows that the reward of such explorations is limited.

There are some fortunate minds whose fertility gives them an ample supply of new ideas. Yet it is human fate that time passes on, and the river of knowledge is mightier than the mightiest single mind. Rarely do the later ideas of even the great possess the relevance of their earlier thoughts. And what of the less fertile minds who, when granted success, tremble in fear of future failure? And what of those intelligent and able seekers on whom fortune has not yet smiled? It is easy to recognize in given instances how luck favored the prepared mind. But the prepared mind is not always favored and no one knows how often the constellation of preparedness and luck occurs.

It is not surprising that many researchers fall by the wayside, some who were not given the revelation of discovery and others who did experience it. There are great artists whose spring of creativity became dry too early and who spent a lifetime being living fossils. There are investigators of renown who outlived by decades the period of their accomplishments. Courageously, some men will by an act of decision terminate early or in middle age their search as investigators. They may weigh the unknown prospects of gain against the certain sacrifices of leisure, breadth, and peace of mind. Coolly, they will compute the probability of future research gains and, judging this probability to be low, devote themselves to teaching alone or to administrative tasks. Often their former colleagues or their younger successors look down on them. "He doesn't do any work any more" is a familiar, cruel comment. But why not permit the honesty of the insight that to create is a hazardous undertaking? He who has tried it has also the right to choose a task where, as a teacher, he can re-create knowledge and attitudes, or where, as an administrator, he can apply his thoughts to the prerequisites of research and teaching. Perhaps he will be able to enjoy the leisure, breadth, and peace of mind whose lack he regretted in his research days. Perhaps he will continue to miss them.

Of course, different men leave research for different reasons. Some give it up reluctantly, but in the knowledge that they can at times be particularly useful in some other capacity. They may fulfill their new duties for a period of time, and then, having carried them to success, return gladly to their original pursuits. Others may even be able to be administrative leaders as well as, simultaneously, investigators. But for these, too, it is valid to say: Carrying out research begifts and deprives.

We tend to admire the man who once having elected a field of study remains with it throughout life. We call him faithful, speak admiringly of his patience, and say that he devoted a lifetime to his specialty. However, what distinguishes his faithfulness from that of a lifelong bookkeeper, his patience from that of a housewife, and his devotion from that of a philatelist? It may be true that the social significance of the scholar's work is greater than that of many other individuals, but is his own personal significance exalted?

Let us therefore be candid with those who want to become investigators. Do not be blinded, we must tell them, by the glamour that is only one aspect of a research career. Be prepared for disappointments and the feelings of failure. But also do not imagine that your own periods of distress are unique. Few of those who seem to have sailed serenely on favorable breezes have not experienced their wind-still periods.

The scientific career has been called a carnivorous god. Perhaps more appropriately it may at times appear a soul-devouring god. By no means, however, does it need to take on this aspect. Whatever dangers personal weaknesses and social pressures may present to the investigator, he can rise above them. He can retain the enthusiasm of youth which led him to contemplate the mysteries of the universe. He can remain grateful for the extraordinary privilege of participating in their exploration. He can incessantly find delight in the discoveries made by other men, those of the past and those of his own times. And he can learn the difficult lesson that the journey itself and not only the great conquest is a fulfillment of human life.

Index of Authors

Note: Italic numbers indicate reference pages.

Index of Subjects

Albinism, 5
Alkaptonuria, 4, 19
Amblycorypha, chromosomal mosaicism in, 62
Anhidrotic ectodermal dysplasia, 103–105
Aniridia, 60
Antibody, 86–89
Ardhanarisvara, 28, 30
Ascaris, gonosomic mosaicism in, 103
Ascertainment, 5
Aspergillus nidulans, mitotic reduction in, 70
Autonomy: in differentiation, 138–141; absence of, 165–167

Barr body, 65, 96
Biometry, 8–15
Birds, gynanders in, 61
Blood groups, A-B-O, 4–5, 13–14
Bombyx mori, double nucleus mosaicism in, 46–48
Brachydactyly, 4, 11
Bufo regularis, and B. carens, prepattern differences in, 162
Butterflies, gynanders in, 31

Carrot, development of, 131
Cat, mosaic in, 70
Cattle, blood mosaicism in, 79–81, 82
Cecydomyidae, gonosomal mosaicism in, 103
Cell lineage, 48–49, 113, 137, 148–149
Chicken, blood mosaicism in, 81
Chromosomes: sex, 18; elimination of, 39–42; somatic crossing over of, 51–56, 89; asynchronous replication of, 98–99

Colias eurytheme, gynander in, 33
Color blindness, 3, 4
Competence, 140–141
Consanguinity, 3, 6
Cytogenetics, human, 17–19

Daucus carota, development of, 131
Deafness, 4
Delphinium ajacis, mutable gene effect in, 168
Developmental fields, 143–145
Differentiation, 130–132; regulation of, 144–145
DNA, 28, 57, 59, 87–89, 98
Dosage compensation, 101–103
Double-nucleus mosaics: in Drosophila, 44–46; in Bombyx mori, 46–48; in man, 73–78; in mice, 78–79; in newts, 79
Down's syndrome, 19, 68–69, 71
Drosophila melanogaster mutants: scute, 134–135, 138–140, 146; achaete, 135, 136–140, 144–145; Hairy-wing, 142–143, 166; Tufted, 142–143; dumpy, 145–146; sexcombless, 151–152; engrailed, 152–154, 168; extra sex comb, 154–159; aristapedia, 158; bithorax, 158; eyeless-dominant, 163–166; multiple wing hairs, 167
Drosophila pseudoobscura, sex determination in, 50–51
Drosophila simulans: inherited mosaicism in, 48; intersexuality in, 90
Drosophila subobscura, cryptic singularities in, 141
Dwarfism, 7
Dzierzon, J., 33